EXPLORERS OF DEEP TIME

EXPLORERS
OF
DEEP TIME

PALEONTOLOGISTS
AND THE HISTORY OF LIFE

ROY PLOTNICK

Columbia University Press

New York

Columbia University Press
Publishers Since 1893
New York   Chichester, West Sussex
cup.columbia.edu

Library of Congress Cataloging-in-Publication Data
Names: Plotnick, Roy E., author.
Title: Explorers of deep time : paleontologists and the history of life / Roy Plotnick.
Description: New York : Columbia University Press, [2021] |
Includes bibliographical references and index.
Identifiers: LCCN 2021027104 (print) | LCCN 2021027105 (ebook) |
ISBN 9780231195348 (hardback) | ISBN 9780231551311 (ebook)
Subjects: LCSH: Paleontology—Methodology. | Paleontologists—Biography.
Classification: LCC QE711.3 .P56 2021 (print) | LCC QE711.3 (ebook) |
DDC 560.1—dc23
LC record available at https://lccn.loc.gov/2021027104
LC ebook record available at https://lccn.loc.gov/2021027105

Cover design: Philip Pascuzzo
Cover images: Alamy and iStockphoto

Dedicated to Arnold and Mildred, who supported me always and in all ways.

CONTENTS

I

DEEP TIME

1

THOSE WHO KNOW THE PAST

ooking out my small office window in Chicago, I see all the
signs of a modern urban environment. Across the nearby
soccer field, a mix of recent and early twentieth-century
buildings are visible (figure 1.1). City buses, trucks, and cars roll
past, dodging pedestrians, and planes fly overhead on their way
to Midway or O'Hare fields. But if I blink and close my eyes, I
can imagine a different scene—some sixteen thousand years ago.
Where I now sit there is a lake, the ancestor of today's Lake Michigan. Looking to the east, I see the edge of a huge glacier that
stretches all the way to what is now Canada. The melting of this
ice sheet has produced the lake in which I am submerged. Toward
the west and south I see the other shores of the lake, ridges and
beaches formed from material left behind as the glacier retreated.
And walking near the shore are mammoths and mastodons and
other animals that will meet their final demise not long after the
ice sheet disappears (figure 1.2).

When I blink and close my eyes again, I have traveled back
more than three hundred million years. I am surrounded by
steaming tropical forests. I see huge trees, but they look nothing
like those of today (figure 1.3). Only the ferns are familiar. I cannot
smell or see flowers. Gigantic dragonflies flit through the leaves,

FIGURE 1.1 Downtown Chicago, Illinois, today, the epitome of the current urban environment.

Source: Photo by the author.

FIGURE 1.2 Illinois in the Pleistocene, sixteen thousand years ago, when mastodons roamed the area.

Source: Mural by Robert Larson, courtesy of the Illinois State Museum.

FIGURE 1.3 Illinois in the Pennsylvanian, about 310 million years ago. Reconstruction at the Field Museum of Natural History.

and an enormous centipede roams underfoot. I see the occasional amphibian or reptile, but there are no birds or mammals. Rivers flow nearby with odd scorpion-like animals and fish swimming in them. Looking southwest, the rivers that course through the forest end in a delta that sits at the edge of a warm ocean.

One more blink takes me back to four hundred million years ago. I am back in the water, but this time I am in a warm tropical ocean that stretches out in all directions. Looking down, I notice that the seafloor is flat and somewhat featureless. Strange animals that look like giant pillbugs walk along the bottom, and large animals that look like squid stuffed into giant ice cream cones swim above them (figure 1.4). In the distance, the waves break on what appears to be a giant coral reef.

Unlike the children in *Willy Wonka and the Chocolate Factory*, I have not entered "a world of pure imagination" when I think about these ancient versions of Chicago. Each of these cerebral trips back into what the writer John McPhee called "deep time"[1] is based on rocks and fossils found throughout the region, all of which I have seen. These samples were collected by me and by the many other scientists who have worked in the area. I am a paleontologist, and these explorations of deep time and the life that lived in the past is

FIGURE 1.4 Illinois in the Silurian. Nautiloids swim above a Silurian reef near what one day will be Chicago.

Source: Diorama at the Milwaukee Public Museum, from the Collection of the Milwaukee Public Museum, #1E013.

what I and my colleagues do. My goal for this book is to explain how we make these explorations, why we do them, and why they are important to society at large. In particular, I want to show you that what paleontologists know about the history of life on Earth is critical for protecting its future.

Paleontology is one of the most familiar and accessible of all sciences. From a young age, children are familiar with the names and life habits of dinosaurs. Media reports of new fossil discoveries appear several times a week. Despite this popularity, misconceptions about paleontology and paleontologists abound: we are confused with archeologists; we study only dinosaurs; or all fossils are rare and valuable. One purpose of this book is to dispel these misconceptions. Another is to describe the "big questions" that drive the field and the methods and data that are used to approach them. I will introduce you to a diverse group of paleontologists and the wide range of topics and interests that motivate them, and you will hear them speak in their own voices as I did when interviewing them on various topics. And I frankly discuss the trials, tribulations, and joys of pursuing a career studying the history of life.

This is not a textbook nor a collector's guide to fossils. You will not be bombarded with a plethora of Latin species names or terms. Instead, I focus on the practice of paleontology, the "how do we find out" part of our science. I also address some of the internal and external controversies of the field, such as the commercial sale of fossils and the value of teaching evolution. Rather than being an "old science," you will learn that paleontology is a young and vibrant field in this day of DNA sequencing and gigantic supercolliders. Incorporating these new technologies, paleontology has shown tremendous relevance for understanding the environmental problems we will face in the future. Finally, I hope I can convince you that paleontologists probably have more fun doing what we do than any other group of scientists.

2

WE HAVE THE BEST QUESTIONS

To make our cerebral trips in time often requires a physical trip in space, sometimes near home, often quite some distance away. A short walk from my home near Chicago is the Forest Home Cemetery, which includes the graves of such disparate notables as the radical Emma Goldman and the evangelist Billy Sunday. Fourteen thousand years ago, the area was the western shore of a lake that would later become Lake Michigan; the lake was formed by meltwater from giant glaciers that covered North America. Along the lakeshore walked a mammoth that died and left its tusk and teeth in a gravel bank; they were uncovered by the cemetery owner in the late nineteenth century. These remains finally ended up in the public library, which is where I discovered them on display. I still find it remarkable and somewhat sad that these magnificent animals lived so near to me in both time and space but I never got to see one alive.

Only a short drive from Chicago I joined the generations of amateur and professional fossil enthusiasts who have collected fossils from the 305-million-year-old Pennsylvanian age rocks of the Mazon Creek region (figure 2.1). Mazon Creek is one of the rare places in the world where remains of soft tissues, not just shells or bones, are preserved. Tens of thousands of hard iron

FIGURE 2.1 Students collecting along the bank of Mazon Creek.
Source: Photo by the author.

carbonate nodules have accumulated here from abandoned strip mines and alongside creeks. Break one of these nodules open and you may be rewarded with a beautiful fern leaf, perhaps a shrimp, or in rare cases a fish, an insect, or a horseshoe crab. Deposited in an ancient delta, the fossils of Mazon Creek are a mix of marine, freshwater, and terrestrial organisms. Although all the species are extinct, the plants and animals of the Mazon Creek are familiar because almost all of them can be placed comfortably in the major groups of life on Earth. But some things are still missing. There are no flowers among the plants from Mazon Creek, there are no butterflies among the insects, and there are no signs of dinosaurs or mammals, although there are amphibians and reptiles.

Also close to Chicago, but some 120 million years older than Mazon Creek, is the huge Thornton Quarry. What makes Thornton Quarry remarkable is that its Silurian age rocks were formed by a large ancient reef similar to those in today's tropical oceans. For those of us living in modern Chicago, it is almost unthinkable that the area that would become our city was once located near the equator, in a climate similar to that of the Bahamas. Although the quarry has abundant fossils of marine organisms, they are not the most attractive of fossils; their shells have been dissolved away, leaving only their imprints, and no soft parts are preserved. The reef was built by corals and sponges, but these specimens are unrelated to what we find in modern oceans. There are ancestors of the modern-day *Nautilus* and a huge number of brachiopods, a minor group today but a major group in Paleozoic oceans. We can also find trilobites, a group of arthropods that once teemed in the oceans but have been gone for 250 million years.

To see older rocks, I traveled much farther from home. High in the Canadian Rockies, in British Columbia, is one of the world's most famous fossil sites, the Burgess Shale.[1] Reaching it is not for the casual traveler. The Burgess Shale can only be reached by a 20-kilometer round trip guided hike, gaining some 760 meters in elevation. But it was well worth the effort and the damage to my knees. The scenery is spectacular and so are the fossils. The Burgess Shale is about eighty million years older than the Silurian at Thornton Quarry and is of Middle Cambrian age. Similar to Mazon Creek, it preserves the impressions of soft tissues. Many of the organisms are familiar and belong to major groups of animals found in the oceans today, for example, sponges. Other fossil animals, such as trilobites, are common in rocks elsewhere of Cambrian age and younger but are long extinct. Yet other fossils are clearly related to animals such as trilobites but are also remarkably different from organisms living after the Cambrian, such as those found at Mazon Creek. And finally, there are those that still just don't seem to easily fit anywhere in the tree of life, but we are pretty certain that they are animals. Many other types of organisms are missing; you would look in vain for an insect or a frog or land plants.

Forty-five hundred kilometers to the east of the Burgess Shale, close to the easternmost point of North America, is the Mistaken Point Ecological Reserve on the Avalon Peninsula of Newfoundland. Despite its name, I did not travel to this remote corner of Canada to observe mink or seals or moose; instead, I came to see one of the most famous fossil sites in the world. To reach it, I had to walk for about forty-five minutes in the rain, over treeless and wet

FIGURE 2.2 Getting ready to examine the Ediacaran biota at Mistaken Point, Newfoundland.

Source: Photo by the author.

ground, toward the coast (figure 2.2). Once there, I shed my shoes and put on the soft footwear known as "Bama boots." The boots were there not to protect my feet but to prevent wear and tear on the surface of the rock and the strange and wonderful fossils embedded within it, the remains of organisms that paleontologists call the Ediacaran biota. What makes these fossils so worth the journey? To begin with, they are almost inconceivably old, dating to about 570 million years ago, some 60 million years older than the Burgess Shale. They are made only of soft tissues, lacking any indication of shells or teeth or bones. There are no signs of organs or appendages, such as eyes or mouths or legs. None of them dug or burrowed into the seafloor; they all lived at or very near the surface, which was covered by a mat made by bacteria and other single-celled organisms. We can squint and think we see resemblances to animals living today, but what is most important is that we really don't know what the Ediacaran fossils are that lived at Mistaken Point. Although many paleontologists would agree that most are related to the oldest true animals, others would place them with lichens, with amoebas, or even suggest that they were a totally independent and now extinct evolution of multicellular life. By the way, as tempting as they are, I did not try to collect the fossils at Mistaken Point or at the Burgess Shale. These sites are carefully protected by the Canadian government, and I was in no mood to be arrested.

What did I learn as I traveled through space and time? The farther back in time I went, the less familiar the living world looked. We no longer have mammoths. The plants and animals of Mazon Creek and the Thornton Quarry include somewhat familiar forms, but they lack many we would find today and contain many that are no longer found on Earth. The Burgess Shale and Mistaken Point are even more alien. So the most important thing I have learned is that *life has a history*. It is a history recorded by the fossils found throughout the world. It is a history that goes unimaginably far back in time, almost to the beginning of the Earth itself, almost 4.5 billion years ago. And it is a history that is revealed and interpreted by paleontologists. We paleontologists are the "explorers of Deep Time," and we use our abilities to reconstruct the life of past ages and the worlds living things occupied.

Recognizing that life has a history is only the beginning. Questions big and small clamor to be answered. Many can be phrased as the simple "what, when, where, how?" When did the mammoths disappear and were humans responsible? Where are there other Silurian reefs? If there were no flowers at Mazon Creek, what did the insects eat? How did recognizable animals appear during the sixty million years between Mistaken Point and the Burgess Shale? And so on. Every question we might ask about the living world today can be asked about life in the past. Paleontology tells us how the life of today came to be and perhaps how it will end. Investigating places such as Mistaken Point and the Burgess Shale give us insights into when the first animals appeared. Looking at the animals themselves tells us what these animals were like and how they compare to forms living today, including ourselves. Putting them into an environmental context indicates where in the oceans these early animals may have evolved.

Recognizing that life has a history raises fundamental questions about that history, questions the late Stephen Jay Gould called the three "eternal metaphors" of paleontology.[2] First, does the history of life have a definite direction—is there an "arrow" of history? Note that directionality does not imply a goal; there is not an endpoint to which the history of life is being driven. Second, how are life and the physical world related—what is the "motor" of organic change? The key issue here is whether life has its own internal dynamic that produces change or whether change is driven by alterations of the external world. The asteroid impact that ended the Cretaceous is an example of the latter type of motor, whereas the evolution of flight among many groups could represent the former. And third, is the history of life gradual or episodic—what is the "tempo" of organic change? For example, the model of "punctuated equilibria," developed by Gould and his colleague Niles Eldredge, suggests that most change within a species occurs in the geologically short period right after it first appears, with relatively little change during the rest of its duration.[3] The aftermath of the proposed impact is another example of a relatively rapid change. In contrast, during a period from about 1.8 billion to 0.8 billion years ago, so little seems to have changed in a biosphere dominated by single-celled organisms that some have called it "the boring billion."

The issues of direction, motor, and tempo are "big questions" that underlie much of paleontological research. These questions can and must be broadened to cover the history not only of life on Earth but the biosphere of the entire Earth system. Is there a directionality to the history of our planet? The number of types of minerals found on Earth has greatly increased over time, and this is due primarily to the presence of life.[4] What drives the long-term changes we see in the history of the Earth and its component systems? For instance, how does the evolution of life affect the atmosphere, or how does continental movement change the distribution of plants and animals? It is now accepted that the Earth has undergone ice ages many times in the past; active research is examining both the possible role of life in triggering glaciations and how life was changed by them. Has the Earth changed gradually over time, or have periodic catastrophes reshaped the face of the globe? It has been suggested, for example, that a huge influx of meteoroids early in Earth's history may have wiped out any early organisms. The work of paleontologists is essential at nearly every stage in the process of unraveling the history of our planet.

We need not restrict ourselves to this planet, however. The active field of astrobiology asks whether life could exist on other planets and what it might be like. Because we know the history of life on only one planet, the insights from paleontology are key to predicting what we might find elsewhere. And it is not unlikely that on planets like Mars, which is currently hostile to life, a fossil record may exist of now extinct biotas.

A question related to that of the motor of change was also raised by Stephen Gould in his book *Wonderful Life*. Gould was concerned about the role of chance in the history of life. To simplify Gould's view, if we were to travel back in time to the Cambrian, we could not predict that the relatively rare members of our own group, the chordates, would rise to prominence, or that the highly abundant trilobites would wane and eventually disappear. The following history is a result of many chance events, each of which may have changed the path of evolution in a different direction. If the asteroid had not struck Earth, would the dinosaurs still rule? Gould's concept of history, which he termed *contingency*, has also been strongly opposed by the paleontologist Simon Conway Morris, who has insisted that much of life's history has been inevitable and predictable.[5] Key evidence for this is the repeated presence of similar forms among unrelated organisms, such as plesiosaurs and dolphins, a phenomenon known as *convergence*. David Raup asked a similar question in the title of his book, *Extinction: Bad Genes or Bad Luck?* The relative roles of chance and predictability in the evolution of life are big questions that are implicit in many areas of paleontology and evolutionary biology. Again, this question can be expanded to the entire Earth system and by extension toward other planets as well. Is life, in particular intelligent life, inevitable in the universe? We know of only one planet on which life has evolved, and the study of how it happened here is the only way we have to directly approach this question.

Paleontology also answers questions for many other fields of science. Within the geosciences, the appearance and disappearance of fossil groups is still the most used method to determine the age of rocks. The abundances and shell chemistries of different ancient organisms are key indicators of ancient climates and changes of ocean chemistry and circulation. The advent of different groups in the fossil record are important constraints on our ability to determine ancient oxygen levels, whereas the pores on fossil leaves provide a check for models of carbon dioxide in the atmosphere. Reconstructions of the former positions of the continents and other geographic features heavily rely on fossil data. Some fossils can tell us how deep in the Earth a rock has been buried and whether it will yield petroleum or natural gas.

Fossils are equally important to and integrated within the biological sciences. Nearly all studies in modern biology are ahistorical, looking only at the current snapshot in time. To truly understand biology is to recognize the role of history, which is recorded by paleontology. The patterns of life we see today are the consequence of evolutionary history, and the fossil record is the direct documentation of that history. The discovery of ancient ancestors and relatives fill in the gaps among living organisms that might look quite different today. For example, comparison of modern forms—based on both their form (morphology) and genetics—shows that whales are related to modern even-toed ungulate mammals, or artiodactyls, that include such familiar animals as pigs, camels, hippopotamuses, and cattle. How one group of artiodactyls evolved into whales can only be understood by looking at the well-documented fossil record of that transition. The fossils also let us date when that split happened. Similarly, fossils are the only direct way to understand how one group of dinosaurs gave rise to birds, or why no modern birds have teeth, although their ancestors did.

One great recent success of paleontology is its integration with the field of evolutionary developmental biology, or "evo-devo." Evo-devo looks at the genetic processes underlying the development of organisms from fertilized egg to adult. As wonderfully documented in Neil Shubin's book, *Your Inner Fish*, the steps in the development of body parts like limbs and which genes are active in each step can be tied directly to changes observed in the fossil record.[6] The fossil record and the study of genetics and development mutually illuminate each other, such as in studying the origin of animals in the Cambrian, the so-called Cambrian explosion.

As one of my colleagues puts it, "We have the best questions!" Topics such as the Cambrian explosion and the extinction of the dinosaurs have caught the imagination of many scientists who are not paleontologists. Scarcely a month goes by without a paper suggesting that some terrestrial or cosmic event "triggered the Cambrian explosion" or "wiped out the mammoths." Some of these suggestions are reasonable and deserve further attention; others just prompt an eye roll and the wish that a paleontologist had been consulted. It is nice that our colleagues are interested in what we do, but they would not have done the work had we not put the questions out there.

Finally, an outstanding issue is how can we best interest our children in and teach them about science? Anyone who has spent time with a seven-year-old realizes that they know the names of all the dinosaurs, what they ate, and how we know that. We can use this knowledge of paleontology to teach concepts of chemistry, physics, and biology, springboarding on the natural enthusiasm children have for fossils.

In sum, paleontology is not—in a phrase that always makes us angry—"stamp collecting." Although describing ancient life forms is essential to what we do, what really drives paleontological research are these big questions: How life on Earth came to be, and how it has changed over time. And one of the biggest "big questions" science can ask—how life will change in the future. We are increasingly interested in virtually traveling forward in time to predict how the world and its biota will change in the face of rapid environmental change.

In the following chapters, I return to these and other big questions and introduce you to many of the scientists who make answering them their life's work and why they decided to do so. I also describe the unique contributions paleontologists make to the biological and Earth sciences. Finally, I hope to convince you that paleontological research is not outdated and archaic. It is a cutting-edge discipline that uses modern methods to answer fundamental questions that cut across the sciences.

3

I'M NOT ROSS (OR INDIANA JONES)

I was sitting with some colleagues at a hotel bar, and the chatty bartender asked what it was we did. Knowing what would come next, we responded with a guarded, "We're paleontologists."

"Cool!!!! Just like Ross on *Friends*!"

Telling someone that you are a paleontologist leads to a fairly predictable set of responses. The most gratifying initial one is "Cool!" The next one is somewhat generational; an eight-year-old will ask if you have found many dinosaurs and a forty-year-old will immediately reference Ross on *Friends* (an alternative is Indiana Jones) and *then* ask if you have found any dinosaurs. This is often followed by either "I want to be a paleontologist!" (the eight-year-old) or "I always wanted to be a paleontologist!" (the forty-year-old). You may also be asked if you go on many digs and have discovered any gold jewelry or pots.

After suppressing a scream, most paleontologists will then calmly explain that going on digs to uncover gold jewelry and pots is archaeology, not paleontology; that Indiana Jones was an archaeologist and a bad one at that (theft of cultural heritage, anyone?); and that Ross was an idiot who would never have been hired as a paleontologist at even the lowest ranked college, let alone get tenure. Most of us will then somewhat sheepishly explain that we also love them, but we have never found, let alone collected, a dinosaur (the eight-year-old immediately loses interest). We would further explain that our research was on some group of organisms that most people had never heard of and was definitely not the stuff of blockbuster movies, although *we* found it exciting.

A safer answer at the beginning would have to been to say, "We're geologists." For much of the history of the science of paleontology, this would have been an equally correct answer. Fossils are found in rocks and serve the important geological purposes of telling us how old the rocks are and in what environment they were formed. Until recently, the majority of paleontologists were trained in university geology departments, and most found work in the oil and coal industries. Paleontology was thus a subdiscipline of geology, or as one nineteenth-century writer disparagingly put it, "paleontology is the handmaiden of geology." Even today, training in earth science is fundamental to the background of every paleontologist, and most paleontology is taught in geology departments.

Another possible answer would have been, "We're biologists." Fossils may be found in rocks, but they are the remains of living things and are not fundamentally different from organisms alive today. Fossils are the best evidence we have for the long-term processes of evolution. Paleontologists must be well trained in the biological sciences, including zoology or botany, ecology, and evolutionary biology. Many paleontologists study both fossil organisms and their living relatives, with information from one illuminating our understanding of the other. Some paleontologists, in fact, prefer to call themselves *paleobiologists*, leaving paleontologist for their more geologically oriented colleagues. A number of these researchers rarely look at fossils themselves, being more concerned with theoretical and statistical analyses of the fossil record.

But when we said "we're paleontologists," what we were really saying is that we are both geologists *and* biologists. To us, fossils are simultaneously things we find in rocks and the remains of living things; thus, they are of geological interest yet are fundamentally biological. In a scientific world increasingly dominated by narrow specializations, paleontology is an incredibly multidisciplinary and interdisciplinary field. A paleontologist can discuss the age of a rock unit like a geologist and the developmental biology of arthropods like a biologist. And many of us are also at home in oceanography, chemistry, engineering, climate science, and a host of other fields. I can think of no other field of science that is so eclectic and multidimensional. And yes, it is "cool."

If paleontology is the study of the history of life, then a paleontologist is a student of that history. Because that history is more than three billion years long and life encompasses everything from bacteria to whales to the entire biosphere, paleontologists usually specialize. These specializations reflect the kinds of fossils they study and why they study them. The names given to these specializations are not simply labels; they can define distinct and sometimes mutually suspicious groups who often have their own journals, meetings, and societies. At the same time, they artificially divide into classes what is in reality a spectrum of interests.

You can identify many paleontologists first by the part of the tree of life in which they are interested. Scientists who study animals past (and often present) with a backbone are vertebrate paleontologists. Vertebrate paleontologists might further self-identify with divisions within the vertebrates, such mammalian or dinosaur paleontologists. Informally someone may be called a "dinosaur person" or a "fish person." These scientists are what the public most often thinks of as paleontologists, with the obligatory beard and cowboy hat (more on this stereotype later).

In nomenclatural contrast to vertebrate paleontologists are the invertebrate paleontologists; their interests include all the other groups of animals on Earth, from snails to sand dollars to shrimp to corals. Invertebrate fossils are far more common and, in many ways, far more important in the overall history of life. But they don't include humans so are naturally less interesting to most people. Would you rather collect a fossil clam or a mammoth tooth? (Thought so.) Invertebrate paleontologists also subdivide along taxonomic lines; much of my early career was as an arthropod paleontologist, which I guess made me an "arthropod guy." The number of people who study particular groups of invertebrates roughly maps a combination of commonness in the fossil record and appeal of the group. A lot of paleontologists study trilobites and brachiopods, but almost no one is looking at fossil annelid worms. For a while, I was one of only three scientists in the world who studied the extinct sea scorpions (eurypterids), although they include the largest arthropods of all time. There are just not a lot of eurypterid fossils. The relationship between commonness and number of researchers does not hold for vertebrates. There are a lot more dinosaur than fish paleontologists, but fish are far more common. Dinosaurs are just the "charismatic megafauna" of the fossil world.

Scientists who study fossil plants are *not* plant paleontologists and would look at you with suspicion if you called them that. They are *paleobotanists*. First cousin to them are the palynologists, whose subject matter is fossil pollen and spores. The plant fossil record is critical to understanding both the ancient and the more recent climate history. Many palynologists work closely with ecologists attempting to understand the history of plant communities since the last ice age, partly in the hope of understanding future changes of plant distribution in a warming world. To be honest, plants get short shrift in many introductory paleontology classes, including mine. I don't have the time, expertise, or samples to cover them. Wish I did.

One somewhat odd sounding specialization is micropaleontology. Technically, this includes the study of any fossil that one needs a microscope to examine (microfossils) and thus includes a hodgepodge of unrelated life forms, including single-celled organisms, plant pollen, and the tiny vertebrate teeth known as conodonts. Jere Lipps, emeritus professor of paleontology at Berkeley, says that this is a "senseless definition if we are truly interested in organisms and their fossil record. We should not talk about microfossils but rather phylogenetic groups." In other words, instead of micropaleontologists, we have paleontologists who study various eukaryotic (protists) and prokaryotic single-celled organisms. Some protist groups made beautiful shells of calcium carbonate or silica and are incredibly abundant in the oceans and lakes of the world. Their preserved shells are fundamental in the study of the longtime history of the world's oceans and climates. They are important clues to what will happen in the warmer and more acid oceans of the near future. These fossils are also key in determining the age of sediments.

The single-celled prokaryotes, bacteria and archaea, are found everywhere in the Earth, including in subsurface and extreme environments. They were also the first living things to appear, so their study is essential to understanding the origins of life. Many of these paleontologists also call themselves geobiologists and work closely with colleagues who study modern representatives of these groups. For obvious reasons, a number are directly involved in the search for life elsewhere in the universe.

Even further down the size scale, molecular paleontologists look at the history of key organic molecules, including DNA. These paleontologists overlap with geochemists who study the history of organic molecules and the environmental signals they cover. Although Jurassic Park is a fantasy, there have been some spectacular finds of ancient DNA and other molecules. The genome of mammoths and Neanderthals have been reconstructed, allowing researchers to better reconstruct the biology of ancient animals and make surprising discoveries about human history.

Another way to identify some paleontologists is by the time period on which they focus. Some are interested in the great explosion of life forms in the Cambrian, others in the terrible extinction in the Permian, or the appearance of modern mammal groups in the Paleocene. Traditionally, a paleontologist was pigeonholed by both age *and* group; when asked, you might say you worked on Ordovician brachiopods or Devonian fishes or Miocene insects. These paleontologists know all the details of the fossil record of their group in that time. Unfortunately, these "old-school" scientists are becoming rarer and rarer, to the detriment of the field. Other ways of doing paleontology have become more popular, and the total number of paleontologists has declined (more on that later). Some types of fossil organisms no longer have a specialist, at least not one in the United States.

The disciplinary jungle becomes even deeper when we include scientists who use fossils as critical pieces of data in their research but may not necessarily consider themselves paleontologists. Fossils are one of the most important tools for reconstructing the environmental history of the Earth, and they provide the context for understanding current and future environmental change. Paleooceanographers use fossils, most often microfossils, to reconstruct ancient ocean temperatures, chemistry, and currents. Paleoclimatologists track the history of ancient climates using a variety of fossil types. The positions of ancient continents and other geographic features are reconstructed using fossils and other geological data by paleogeographers. Related to these are paleoecologists, who generally focus on changes in plant and animal communities over the past one hundred thousand years. And if you are not confused yet, another kind of paleoecologist uses the distribution of fossils to reconstruct ancient environments, such as deltas or reefs.

Fossils also play a critical role in understanding the evolution of life on Earth, and many paleontologists also call themselves *evolutionary biologists*. Extinct groups of organisms fill out the tree of life, clarifying relationships and origins of species alive today, including our own. Fossils also act as critical calibration points in the numerous studies that attempt to determine when modern groups originated. For example, any attempt to determine when birds originated, based on their DNA or RNA, must accommodate *Archaeopteryx*. The study of evolutionary transitions in morphology, such as the tetrapod limb, are now integrated with genetic and molecular studies of embryonic development, the science of evolutionary developmental biology, or *evo-devo*. This integration was wonderfully explained in Neil Shubin's book *Your Inner Fish*. Shubin (as is clear from the book) is an excellent example of a paleontologist who slides from geology to biology and back again.

Yet another way to parse the community of paleontologists is by where they work. I discuss this in more detail in later chapters. Museums of natural history are probably the first places that come to mind. The image of the paleontologist carefully putting dinosaur bones (of course) together has been used frequently in movies, television shows, and comics, perhaps most famously by Cary Grant in *Bringing Up Baby*. Major natural history museums, such as the American Museum of Natural History, the National Museum of Natural History, and the Field Museum in Chicago, all employ multiple paleontologists, usually from several disciplines.

In addition to the large natural history museums, many smaller museums also employ paleontologists. These may be state museums, museums attached to major universities, or local museums such as the Burpee Museum in Rockford, Illinois. These museums may have only one or two paleontologists and a few staff. Others, such as the Yale Peabody Museum, share faculty with university departments and can be quite large.

Paleontologists are also employed at universities and colleges. Usually, they are in geology or earth science departments. Some are in biology departments, and more than a few vertebrate paleontologists teach at medical schools. As biology departments become increasingly dominated by cellular and molecular biologists, they have become dependent on skilled vertebrate paleontologists to teach such core courses as comparative anatomy. Except for a few very large programs, paleontology professors also teach other subjects. My own teaching responsibilities, for example, included introductory earth science, statistical methods in earth and environmental science, Earth history, as well as paleontology. I also supervised graduate students, served on committees, interacted with the public, and conducted my own research.

Governments provide another niche for paleontologists, where they are part of a large coterie of scientists responsible for evaluating our natural resources for the common good. At the state level, they can be employees of the state geological or natural resource surveys. At the federal level, paleontologists work for an alphabet soup of agencies, including the National Park Service (NPS), the National Forest Service (NFS), and the Bureau of Land Management (BLM). As you will see, these individuals are at the forefront of efforts to protect fossils on federal land while making them available for research, which has led to controversies among their paleontological colleagues.

Related to these are the growing ranks of "mitigation paleontologists," who collect fossils in front of construction equipment, saving them from destruction. Much of their work is done under contract in compliance with state laws similar to those that protect archeological sites. Although some mitigation paleontologists have advanced degrees, it is also possible to find employment with a bachelor's degree.[1]

For many years, most paleontologists worked in the fossil fuel industry. They mainly worked on various microfossil groups and received the sobriquet "bug pickers" or, more formally, economic paleontologists. This now archaic usage is memorialized in the original name of the association of geologists who study sediments and sedimentary rocks: Society of Economic Paleontologists and Mineralogists. The bug pickers did, and some cases still do, provide critical information on the age of the rocks and environments, which was essential for constraining the conditions for the formation and preservation of fossil fuel. The once large number of economic paleontologists has shrunk dramatically in recent years; an article published in 2000 indicated a 90 percent drop in oil company employment since 1985.[2] (I do not know the current numbers.) Industry support for academic paleontology has fallen in sync with this. However, some training in paleontology is still considered important for petroleum geologists.

Another group to mention here are those individuals who make their living collecting, preparing, and selling fossils. The very existence of these commercial collectors makes some of my colleagues, in particular those in vertebrate paleontology, bare their teeth and snarl. Others, including myself, are more tolerant. (I return to this issue later in the book.)

Finally, some people collect and study fossils as a hobby and a passion. These amateur and avocational paleontologists have made key discoveries in the history of our field and are key companions in paleontological research.

In sum, we paleontologists are anything but monolithic. We encompass a wide range of interests, skills, and ways of making a living, but when compared to other fields of science, we are tiny. In 2008 I estimated that there were on the order of one thousand academic paleontologists in the United States.[3] This is much fewer than the number of new physics PhDs awarded every year. An advantage of our limited numbers is that paleontology, to a large extent, is a family. Many of us know each other and are related by shared colleges and grad schools. Academically, I have paleontologist brothers, sisters, and cousins, and in a few cases uncles and aunts. Like any other family, there are jealousies and quarrels, some petty, some substantive. I have heard colleagues called crackpots to their faces. And one or two are, quite frankly, embarrassing to the rest of us. Even so, there is true warmth and affection within the community. We respect each other's work, and when we disagree, we usually do so politely. When we find a colleague's results and data exciting, we let them know. We look for opportunities to share and collaborate. And some of us have even found dinosaurs.

II

EXPLORING
DEEP TIME

4

ATTENDING MARVELS

Sometimes you just get lucky. In 2004 I was trying to decide where to take my advanced class in paleontology for a field trip. This was not easy. Nearly all the of the landscape near Chicago is covered by sediments left behind by the great glaciers of the last ice age. Looking for fossils usually involves visiting one of the many rock quarries that have been dug through the glacial material to reach the bedrock beneath. One such limestone quarry was described in a publication of the Illinois State Geological Survey. Only an hour and a half drive away, it was excavated in fossiliferous rocks of the Ordovician age. Perfect!

My small group visited the quarry on a pleasant Saturday. Looking at the south wall, we quickly realized something didn't look right. Embedded within the expected tan and gray limestones of the Ordovician were masses of sandstone, thin coals, and shale, which contained abundant fossil plant fragments. These rocks obviously came from the much younger Pennsylvanian, usually found miles to the south. Then we noticed that the Ordovician limestones surrounded the Pennsylvanian sediments not only on the bottoms and side but on the top. One of the graduate students, Todd Ventura, then realized what we were seeing: "It's a cave!"

FIGURE 4.1 310-million-year-old cave in Central Quarry, Illinois.
Source: Photo by the author.

And so it was. What we had discovered were fossil caves or paleokarsts (figure 4.1). They formed about 310 million years ago, when an ice age in the Southern Hemisphere dropped global sea levels and exposed limestones and dolomites that had formed hundreds of millions of years earlier. Rain falling on these ancient carbonates produced caves, just as they are doing in Kentucky at Mammoth Cave today. Shortly afterward, rivers coming from the northeast filled and covered these caves, which prevented them from collapsing and aided in their long-term preservation.

This was my first of many visits to that site and similar ones elsewhere in Illinois, accompanied by students and my UIC colleague Fabien Kenig and paleobotanists Andrew Scott and Ian Glasspool. These caves proved to be a treasure trove for paleontology and geology. We could establish that the caves and the sediments that filled them were about the same age. We identified an adjacent fault that may have been at least partly responsible for the cave formation. And we found lots and lots of fossil plant remains, including leaves and spores, many of which had been burned and turned to charcoal prior to washing into the cave. The plant fossils included some of the oldest conifers in the world. A close look at the leaves with a scanning electron microscope (SEM) showed beautifully preserved stomata, the pores through which leaves breathe. The spores preserved the oldest evidence of the chemical sporopollenin, which makes up their resistant outer walls. Not only had the plants been burned, but preserved chemicals indicated that it was a very hot fire. The only animals we found were tiny scraps of scorpions. But even these were amazing because they preserved some of the oldest evidence we have for the chitin and protein that makes up arthropod exoskeletons.[1]

Our cave research has some similarities with many other field-based research projects. First, finding fossils is a combination of knowing where to look and luck. Second, fieldwork can be rewarding but is also stressful and frustrating. One of our return visits occurred on a 95°F (35°C) day, when the limestone walls and floors reflected the heat and baked us. Not fun. And about five years after we found the site, the family that had owned the quarry for eighty years sold it. The new owner wouldn't let me in, nor return my phone calls and emails. We could still visit two other sites but could not get close to the quarry walls due to safety concerns passed down from a federal agency. Third, fieldwork was just the beginning. There was a tremendous amount of work involved in processing the materials we collected to figure out what we had. Fourth, a lot of the analyses involved the high-tech analytical equipment that is now standard in our field, including SEMs, mass spectrometers, and something called an X-ray absorption near edge structure (XANES) spectrometer. Fifth, students participated, not only to do the heavy work but to learn about paleontological research in the field. Finally, it has been a great excuse for me to learn about new stuff and to explore new areas of research. Before I found the paleokarst, I did not even know that such a thing existed. I knew very little about caves or the fossils we find in them. The geological and paleontological history of caves became a major area of my research.

One of the most famous cases of "fortune favoring the prepared mind" is that of Charles D. Walcott, an invertebrate paleontologist and secretary of the Smithsonian Institution. Beginning in 1907, Walcott spent his summers doing fieldwork in the Canadian Rockies. His goal was to describe the Cambrian rocks in the area, many of which contained spectacular trilobite fossils. Legend has it that in 1908 his wife's horse kicked over a loose piece of rock that contained a fossil he had never seen before. True or not, he recognized that the loose pieces of rock along the trail contained a remarkable collection of fossils, where impressions of the original soft parts were preserved. Rocks only move downhill, so he knew that the source of these specimens was up the slope from him. Returning the following year, he opened a quarry to systematically excavate these fossils. This was the discovery of the Burgess Shale, one of most significant and famous fossil deposits in the world. Until recently, this had been our best glimpse of the remarkable animal life of the Cambrian, some 500 million years ago.

Another serendipitous discovery was made by Johnny Waters (Appalachian State University), who has been doing fieldwork for forty years all over the world, including on "the border between Ireland and Northern Ireland during The Troubles, when we were stopped by a British Army patrol; the Xinjiang Autonomous Region, China, where the Uyghur concentration camps are; including one trip in 2000 when we were deported for being too close to the border with Kazakhstan; . . . and Yunnan, where we were doing fieldwork in the midst of a swine flu outbreak." In 2012, Waters was doing fieldwork in Mongolia. They were visiting with a group of Mongolian geologists when they learned "that the cook van had overturned. The brakes had failed as the driver attempted to climb up a hill." After much effort, they got the van upright and "did what geologists always do—we looked at the rocks." The site, which was named the "Hushoot Shiveetiin gol, or the car wreck locality," became a very important locality for understanding Middle Paleozoic climate and biodiversity change, was the site of a large field workshop, and has resulted in six publications to date.

There are very few disciplines that allow you to combine a love of science with a love of the outdoors. Most of geology, of which paleontology is a part, is one. Another is ecology, which is close in many ways to paleontology. And there is archaeology, which is almost, but not quite, the same thing as paleontology (there are many shared concepts and methods). For many paleontologists, "the field," our term for the places we explore outdoors, is where we are happiest. Most of our stories are field trip stories. A new field area is for the paleontologist as a powerful new telescope is for an astronomer—it has the potential for new discoveries that will change how we look at life's history. As Ellen Currano (University of Wyoming) says, "Field paleontology is adventure with a purpose." Jacques Gauthier (Yale University) simply says that "fieldwork is essential to my well-being!" Although I do not consider myself a field paleontologist, I get itchy if I don't get out occasionally to bang on some rocks.

One of the questions often asked of a paleontologist is, "How do you know where to look?" A simplistic answer might be: "I know where rocks of the right age and environment are to find the fossils I am looking for." This begs the follow-up question: "How do you know where those rocks are?" A frequent response to this question is, "Someone else described those rocks and I read that description." What this really means is that paleontology is an old science (no pun intended); the systematic study of rocks and their fossils has been going on for about two hundred years. As a result, many places, especially in Europe and North America, have had a geologist or paleontologist visit them at some point.

Sitting in my office are seventeen huge volumes dating from as far back as 1842 to about 1890. These comprise the *Geology of New York* and the *Paleontology of New York*, which were produced by the state-supported New York State Geological Survey. The latter set is mostly the work of James Hall, the New York state paleontologist. In these volumes are descriptions of fossils and localities throughout New York. I also own nineteenth- and early-twentieth-century volumes from the geological surveys of Ohio and Illinois, and from the United States Geological Survey (USGS). The mission of these surveys was, and still is, to document the mineral, water, and other resources of their states and the country. Similar organizations exist throughout the world.

One way geological surveys fulfill their mission is to map local geology, often in incredible detail. These surveys are published in various formats, which include descriptions of the local geology, maps, and a list of fossils. Another group that maps local geology and sometimes publishes it publicly are oil and gas company geologists, who explore remote areas in search of undiscovered resources. Nearly all of these works can be consulted in college and university libraries, and many are now available online.

Also to be found in libraries are the thousands of master's and doctoral theses that have been written about local geology and the scientific papers that followed. For decades, most geology and paleontology graduate degrees involved the detailed description of a small area of the Earth's vast surface, including the fossils that were found. After these researchers received their degrees, they often continued to write papers on the fossils of local areas. These papers are the critical underlying data of paleontology.

Much of this information is summarized in various textbooks used to teach the next generation of geologists and paleontologists. Neil Shubin, in

Your Inner Fish, describes how they located the rocks that eventually yielded the early tetrapod *Tiktaalik*, a key intermediate in the transition from fish to terrestrial vertebrates. Based on earlier work in Pennsylvania, he and his colleague Ted Daeschler knew they wanted rocks of Devonian age that formed in freshwater environments. By chance, they found a diagram in an introductory historical geology textbook that showed rocks of the right age and environment in the Canadian Arctic. They knew where they wanted to go and eventually went. In sum, the first step in finding fossils is to dig—not in the rocks but in the library.

New technology has become a valuable aid in planning. Satellite images available on Google Earth and elsewhere, online topographic and geologic maps available from the USGS, and geographic information system overlays such as land ownership are examined in detail before the first step out the door. One of my favorite websites is ACME Mapper, which allows easy toggling among satellite images, road maps, and topographic maps.[2] Another is TopoView from the USGS, which includes the ability to look at historical topographic maps in an area.[3] This is quite valuable if you would like to reinvestigate a site described one hundred years ago.

What we collect and how we collect is a function of the scientific questions we are asking. Some of these questions are fundamentally geological: What is the age of the rocks? In what environment did they form? For example, Michał Kowalewski (University of Florida) and Susan Kidwell (University of Chicago; box 4.1) have long been interested in how much time a "shell bed" represents.[4] Shell beds are rock units that, as the name suggests, are densely packed with shells. Understanding the relevant geologic processes has led to the realization that shell beds can contain individuals that lived hundreds of years apart. Many fossil collections, therefore, represent an average of conditions over long periods of time rather than any instant in time.

BOX 4.1

Susan Kidwell describes herself as "a geologist who works on fossils." Susan is originally from Virginia. As a child, she loved being outdoors and loved natural history. At first, she wanted to be a park ranger, but her romantic notions about the job were disabused after reading Edward Abbey's Desert Solitaire. *She then realized that the people who did the science in the parks tended to be professors and others from outside. Susan's father worked for the United States Geological Survey, and she learned from the geologists there that the job involved being outdoors and that let you do "cool things." She knew she wanted to be a geology major and to get a PhD when she entered college at William & Mary. Susan went on to get both a master's degree and a PhD at Yale University.*

Other questions are more biological: Are new fossils present that fill in gaps in our understanding of the history of life? What was the ancient ecological community like? Is there evidence for predation or competition? One of the best pieces of evidence we have for predation in the geological past are the distinct drill holes made by predaceous moon snails on their snail and bivalve prey. Patricia Kelley (University of North Carolina Wilmington), along with her students and collaborators, have collected tens of thousands of shells from numerous geological units, with the goal of understanding the history of this predator-prey interaction.[5]

What makes the covers of the journals of *Science* or *Nature* or the front page of the *New York Times* are generally the more spectacular and rare finds. This usually starts with "Paleontologists have discovered the earliest . . ." or "The biggest . . ." or "New relative of *T. rex* . . ." etc. The late Martin Brasier called this the "Mofaotyof principle," which stands for "my oldest fossils are older than your oldest fossils," and represents the excitement, attendant publicity, and as Brasier stresses, the necessity for concrete evidence when the oldest member of a fossil group is first discovered and published.[6] In most cases, however, we are just as or perhaps more interested in the vast majority of common and somewhat mundane fossils that make up the history of life. As Ben Dattilo (Purdue University Fort Wayne; box 4.2) said to me, "Rather than chase after rare fossils, I have taken to studying the fossils as part of the rock record, and I have concentrated on the more common things. Most of my work is done in the Cincinnati, Ohio, area where repeated visits to the same outcrops allow me to observe and question on a level of detail that adds to a body of work that has been done on these rocks for almost two hundred years. For example, I have done ongoing research on the brachiopod *Rafinesquina alternata.* If you cannot find thirty of these in as many minutes at a typical Cincinnatian outcrop, you just aren't trying. It is arguably one of the most common macrofossil species on earth. This means it has been overlooked and passed up for detailed study—because it is not rare."[7] I agree strongly with Ben on this; the vast majority of fossils are *not* rare, although some are. And nearly all field discoveries are important to building the body of knowledge even though they may not land us on the front page of the *New York Times.*

BOX 4.2

Ben Dattilo grew up in Madison, Indiana, halfway between Cincinnati and Louisville. He collected his first fossil when he was four and "continued collecting fossils until my early teens, with The Golden Guide to Fossils *as my principal guide." He was fortunate enough in 1977 to meet a paleontology graduate student who permanently sparked his interest. Nevertheless, when he first went to Brigham Young University, he was being encouraged to be an engineer. While on a Mormon Mission, Ben had a conversation with someone who said something like this: "You do not want to be an engineer, the training just makes people narrow minded." That was all it took. When he returned to BYU, he took a course on the history of life and ended up as a geology major. "Shortly thereafter I started working in the Earth Science Museum where I worked as a vertebrate fossil preparator and collector." After getting his bachelor's and master's degrees at BYU, Ben went on to get his PhD at the University of Cincinnati, writing his thesis on the wonderfully exposed Ordovician-age rocks in that area. He still enthusiastically studies them.*

One of the most dedicated field-oriented invertebrate paleontologists was the late Arthur Boucot. Art was a large and colorful guy known for collecting pretty much everything available at an outcrop when looking for rare fossils; his propensity to do this has given rise to the phrase "to boucotize" an outcrop. I have also heard that having two hundred pounds of rocks on your back is "a boucot." Art's love and encouragement for fieldwork is memorialized by his funding of the Paleontological Society Arthur James Boucot Research Grants, which encourages fieldwork-based research by young scientists.

Another legendary field paleontologist is Carlton Brett of the University of Cincinnati. Those of us who have been privileged to be in the field with him are struck by his knowledge of the geology and the fossils, his sophisticated interpretations of the environments in which the fossils formed, and most of all by his infectious enthusiasm. As Arnie Miller put it: "Carl is, first and foremost, an aficionado of fieldwork and there is no place that he would rather be at any given moment on any given day (or night) in any given weather condition than at a rock outcrop."[8] The "night" part is a bit of an inside joke among paleontologists; Brett is notorious for not wanting to leave the field even when it is dark, so you end up looking at an outcrop under automobile headlights.

When we are doing fossil collecting, we are not necessarily "on a dig." Many times, the fossils we are looking for are right at or near the surface, and very little systematic digging is involved. As Jere Lipps put it to me: "I've been on vertebrate paleontology and invertebrate paleontology digs where the overburden was removed centimeter by centimeter to get to a bone or shell bed or accumulation that was then excavated very carefully with grid systems, photographed, each specimen labeled and carefully removed. *Dig*, when used in this way, seems to me to mean a planned activity that takes some time." Picking up a fossil exposed on an outcrop is not a "dig."

Even if you know where you want to collect, you may not be able to get access. Much of the bedrock geology of the Midwest is covered by materials left by the last ice age. These glacial deposits contain cool stuff including mammoths and mastodons. But for those of us who are interested in the far more ancient fossils contained in the bedrock, these overlying deposits ("the overburden") must be removed before our sites are accessible. In some places, the bedrock is exposed by stream action, but these outcrops tend to be sparse. Then there are roadcuts, which might expose a considerable amount of rock but are sometimes hazardous due to passing traffic or are unavailable due to legal restrictions. The best sites are rock quarries, which until recently were quite welcoming to collectors, both professional and amateur. Fears of lawsuits and complications with federal safely regulators have markedly reduced access. I recently scheduled a paleontology class visit to a quarry in western Illinois. On the Monday before our Saturday excursion, our contact emailed me to say that "the plant regrets a last-minute cancellation but was visited on site by MHSA this past week in a routine inspection. We are forced to withdraw the invitation." MHSA is the federal Mine Health and Safety Administration responsible for ensuring safe conditions at the nation's quarries and mines. I was never able to find out what issues prompted the cancellation. I had to scramble until I could find another quarry willing to accommodate us.

Access to private lands is always an issue. It is vital to build trust with local landowners to overcome any suspicions they may have of why you want access to their property. Some of this is the legacy of the collecting of the *T. rex* known as Sue, which ended up impounded over issues of land ownership and access. One of my former students, who is from Montana, has been very successful in getting access to field sites in that area because she speaks the language of the landowners. Public lands have also become problematic, with increasing restrictions placed on collecting. I return to these issues in chapter 16.

Once you have an idea of where to look, you must gear up. Gear for the field varies from simple to complex, from high tech to low tech or even no tech at all. The simplest tools of all are the trained eye and brain. Much of a paleontologist's time is spent scanning the ground or the cliff face, looking for that elusive bit of color or shape that signals the presence of an item of interest. The best fossil hunters literally seem able to "smell" the fossils. A similar ability among mushroom hunters was described by Michael Pollan in *The Omnivore's Dilemma* as "getting his eyes on."

Fossils, like the living things they come from, tend to be small. Even large bones can shatter into thousands of small fragments. So around the paleontologist's neck is a hand lens, a high-quality magnifying glass that is similar to the loupe used by jewelers. The item of interest is examined closely under the hand lens, leading to the decision to keep it or more likely toss it aside (an old joke used on students is that such specimens are leaverites, as in "leave it right there").

Fossils also tend to be embedded in rock or at least in compact sediments, from which they need to be separated. If the material is soft or loose, an old butter knife is a useful tool. If it is hard, however, it must be broken or split, and for that you need a hammer. Not just any hammer, mind you, but a sturdy steel hammer with a comfortable handle. A flat chisel end is handy for splitting rocks. A well-made hammer, with good balance, is not only easy to use but is fun to bash on rocks. In some cases, if the rock is hard, a large crack hammer or sledgehammer can be used to break it up. One of the chief tasks of a paleontologist in the field with students is to get them to *not* use their hammers at every opportunity. The other task is to get them to put safety glasses on when hammering on hard rocks (splinters will fly). Murphy's Law for field paleontology is that the rock will always split across the specimen you are interested in recovering. The hammer also can be used to smack a chisel into the rocks, to separate the layers or to reduce a large block into something more manageable. Peter Ward has called these "the tools of my trade, the cold steel of hammer and chisel which are my access back into time."[9]

A hand lens, a hammer, a chisel, safety glasses, toilet paper to wrap small or fragile specimens, plastic bags for fossils, a pen to label the bags and make notes, a waterproof notebook to record observations and what is collected, perhaps an old toothbrush or small whisk broom to remove dirt, a tape measure, and a sturdy backpack to put all of this stuff into is the basic field kit. A lot more can be added. If you are working near vertical cliffs, a hardhat is a necessity—it is required by law if you are working in a quarry. A camera is important for recording information about the site; a map and a GPS unit are vital to determine where you are and to accurately record locality data; bug spray is critical in some climates and times of year; and, of course, we always wear sunscreen. Tablet or laptop computers have become standard ways of taking notes. Many of us also pack a small bottle of dilute hydrochloric acid; "dropping acid" tells us if our rocks are limestones. If you are working on relatively recent fossils in sediments that have not yet become rocks, a sieve is useful for finding very small things.

Things get more and more complex the longer you plan to be out in the field and depending on the location of your field site. I am fortunate that my field sites are less than two hours from home. I can drive there and spend the night in my own bed. Many sites, however, require you to spend the night far from home or a pleasant motel (which costs too much anyway). Camping out is standard operating procedure, and in some cases this is far from even the minimal amenities of water and a pit toilet. Of course, this means bringing all the gear needed for camping, as well as food, perhaps water, and any supplies you might need, such as bug spray. For remote field sites such as in the Arctic, this gear must be flown in, at a not inconsiderable expense. As Neil Shubin points out, the logistics of fieldwork in remote areas is complex, requiring "at least eight days planning for every day in the field."

The technological revolution has also transformed paleontological field-work. One of the most important things in the field is knowing exactly where you are. Detailed location information, usually including a detailed map, is part of every paper describing a field site. Traditionally, paleontologists and geologists have used topographic maps and high-end compasses to orient themselves and mark their collecting sites. Learning to read a topographic map, and to use it in the field, was and is part of the basic training of Earth scientists. The advent of global positioning systems (GPS) has changed all that. Once used only by the military, almost every cell phone can now locate you on the surface of the Earth to within 10 meters, and you can purchase GPS equipment that has precision on the centimeter scale. The locations can then be mapped with high precision using geographic information systems (GIS), the same technology that tells you the location of the nearest takeout pizza. The GPS coordinates of a fossil site are now essential pieces of information to be reported.

Everything that is transported in also needs to be transported out. This includes not only gear but the fossils themselves and any material needed to protect them during the journey home. As Anne Weil (Oklahoma State University; figure 4.2, box 4.3) puts it: "This can be very physically demanding since I worked very intensively in wilderness areas from which one has to hike hundreds of pounds of matrix out. . . . I think this is what people really imagine vertebrate paleontology is like, and it is just like that a lot of the time—filthy, sweaty, exhausting, and very satisfying." Susan Butts recalls someone calling this "blood, sweat, and beers." During his 2012 Mongolia expedition, Johnny Waters and his field party "had to literally transport everything it needed—food, water, gas, etc. We traveled in Russian vans on roads that are basically tracks across the desert. . . . The camp had a cooking tent, individual tents for sleeping, and a very cold stream for bathing. There were no bathroom facilities."[10]

Being out in the field often means being away from a convenient toilet, even a pit toilet. For some beginning students, the very idea of going to the bathroom outdoors is problematic, let alone the lack of showers. As Anne Weil explains it: "Solid waste disposal depends entirely on the regulations enacted by the agency or preference of the landowner." Some require burial on site, others that it be hiked out. She continues, "One efficient way to do the latter is to use a "groover," or toilet-dedicated ammo can, so named for

FIGURE 4.2 Anne Weil at the Homestead Site, an Oklahoma fossil site.

Source: Photo by Jeff Hargrave.

BOX 4.3

Anne Weil is a vertebrate paleontologist at the Center for Health Sciences of Oklahoma State University. Anne considers herself primarily a geologist. She came to paleontology somewhat late. She was interested in journalism in high school and began as an English major at Harvard. As part of the breadth requirement, she took a class with Stephen Jay Gould on the history of life. Gould, in turn, suggested that she also take physical geology to complete her requirements. Her instructor urged her to go into geology. Anne did end up an English major but took enough geology to be able to enroll for a master's degree in geology at the University of Texas at Austin. She then went on to study at Berkeley with the great vertebrate paleontologist William Clemens. Like many vertebrate paleontologists, she is employed at a medical school, teaching courses such as gross anatomy. She calls herself "a geologist in a medical school."

the grooves it leaves on one's fleshy hind parts." At the least, many of us are already carrying a roll of toilet paper and hand sanitizer.

This is one area of fieldwork where sex differences really stand out. For men, urination is usually no more complicated than finding a nearby tree, bush, or rock to mark as your territory. For women, it is often more difficult to find a private spot. Then there are additional complications associated with menstrual periods. Anne Weil again: "I think a lot of women are afraid that someone will notice the terrible, shameful event of their period. 'What if someone sees the bag in the trash??!' To which I say, 'So? It's not like it was a mystery to them that you are a female of reproductive age.' " I spent a summer in the field with a nursing mother who needed to express breast milk, while her own mother cared for the child. Marilyn Fox, who leads field trips for the Yale Peabody Museum, has a necessary discussion with undergraduates about "female things in the field." None of this has stopped women paleontologists from being as successful at fieldwork as their male counterparts.

This is assuming that cover is available and the weather is warm. On subfreezing nights, one fights the competing urges to urinate and to stay warm in your sleeping bag. For those working in the polar regions, sometimes the only solution at night or in foul weather is a dedicated bottle. As I tell my students, if you see a toilet, use it. Who knows when there will be another opportunity.

This brings us to hazards. We often work near steep cliffs or quarry walls. Falling on rocks or having rocks fall on you is a real possibility. The rules here are to never stand directly above someone and to loudly yell "rock!" if you knock one loose. If you are outdoors away from shelter, there is the risk of lightning strikes or flash floods. If you are working along a sea cliff, you risk being trapped or drowned if you don't keep an eye on the tide.

Animal bites are another danger; in areas with venomous snakes or scorpions, the basic rule is to turn over rocks with your hammer, not your hand. In Alaska, northern Canada, and many parts of the American West, bear attacks are not a minimal risk. Rifles are typically part of the field gear in those areas. Then there are the mosquitoes and the black flies. George Gaylord Simpson, in his 1934 classic *Attending Marvels*, tells of doing fieldwork in Patagonia in 1930. Not only were there strong and nearly unending winds ("Its brutal force and nerve-destroying continuity almost pass belief") but also millions of flies, as well as scorpions, spiders, and centipedes. He also describes a monotonous diet of mutton: "Dinner which we shared, consisted of boiled mutton, cooked so long that it was literally putrid, followed by mutton broth. For dessert, a bowl of hot mutton grease was brought in and bits of hard tack were soaked in this."[11]

Nearly ninety years later, paleontologists working in Mongolia describe similar experiences, including endless meals based on mutton. Ross Anderson (University of Oxford) simply characterizes the food as "leaving much to be desired," and Pedro Marenco (Bryn Mawr College; box 4.4), a committed carnivore, found that it was too much meat even for him. Marenco also recalls a rather gruesome encounter with a local insect pest. He was working in Mongolia with Stephen Dornbos (then at the University of Wisconsin–Milwaukee) and a team led by Tatsuo Oji from Japan, accompanied by the Mongolian geologist Sersmaa Gonchigdorj. The group was looking for fossils preserving soft tissues in rocks of Ediacaran age, about 555 million years old. When they arrived at the field site, they realized a trench had been dug in front of it that was filled with dead animals, mostly goats and sheep but also included a horse. You can imagine the smell and the swarms of insects. They set up camp. Not long afterward, the camp cook, who was also a doctor, was called over to look at the eye of one of the Japanese students and realized that a fly had laid eggs in it, which had already hatched into maggots. Thankfully, she was able to remove them before there was any permanent damage.

BOX 4.4

Pedro Marenco was born in the United States, but his family did not emigrate permanently from Nicaragua to southern California until the early 1980s. He had been interested in fossils since he was seven years old, when he recorded himself reading kid's books on dinosaurs. As a child, he thought paleontology was just dinosaurs. As is the case with many others of his generation, he grew up with computers and video games and loved programming, producing his first line of code when he was eight years old. Inevitably, he entered the University of Southern California (USC) as a computer engineering and computer science major. During his junior year, Pedro took a course in invertebrate paleontology and evolution with David Bottjer and discovered that paleontology was more than dinosaurs.

He particularly liked the field trips. Pedro contacted Bottjer about a job and became a field assistant to Dave's then graduate student Margaret Fraiser. He became smitten with the science and stayed at USC for a master's and PhD with Bottjer.

They eventually learned that the culprit was the spotted flesh fly, which as the name implies lays its eggs anywhere it can, but especially in exposed soft tissues. It dive bombs toward places like eyes, which is why wearing glasses helps prevent infection. At one point, Marenco felt a thud and found eggs laid on his eyeglasses. The fly is a scourge of livestock, which is probably why the carcasses were in the pit in the first place. Carefully keeping their mouths closed, Steve and Pedro and the rest of the team buried the animals so they could work in some comfort.[12]

When Stephen Dornbos was a graduate student at USC, he accompanied Frank Corsetti and Junyuan Chen (Nanjing Institute of Geology and Palaeontology) on some fieldwork in the Hunan and Guizhou provinces of China, studying rocks from the period of Earth's history before the Ediacaran, informally known as "Snowball Earth."[13]

We wanted to sample this late Precambrian manganese carbonate for geochemistry and micropaleontology because it was potentially a Snowball Earth cap carbonate . . . we visited an active manganese mine to get some good samples. We gathered in the mining office and were issued knee-high rubber boots. Now, being a tall person, I have large feet. The largest boots they had were way too small for me, so I curled up my toes figuring that it wouldn't be much of a walk. I was mistaken. We ended up delving over a kilometer into this active mine that had low (5 feet or so) unsupported ceilings, no lighting, no ventilation, and extremely poor drainage. Hunched over in our ill-fitting boots, we sloshed slowly through knee-high black waters all the way to where they were actively drilling. Members of our party started banging on the walls of the mine with their hammers to get samples as Frank and I peered nervously at the unsupported ceiling. We got our samples and got out of there as fast as we could. We emerged from the mine covered in black sludge and were offered some soap to wash up at a faucet just outside the mine. As we were washing our hands, we observed that the pipe leading to this faucet headed straight back into the mine. We asked for even more soap. I think of this experience whenever I hear about too frequent fatal Chinese mine accidents.

Ellen Currano (box 4.5) is a paleobotanist focused on determining how ancient forests respond to environmental change. Here she describes some of her fieldwork experiences.

Fieldwork is my favorite part of my job—I love wandering the badlands, imagining what they would have looked like when the rocks were forming, and collecting new fossils that will provide new information. . . . Paleobotanical fieldwork really is not glamorous. I spend a lot of time walking the badlands, digging holes in the ground looking for leaf fossil localities, and then when we find good ones, we dig really big holes, carefully pry out blocks of fossiliferous rock, and wrap them in toilet paper. I've sure sweated it out in the Wyoming badlands and Kenya's Turkana Basin (midday temperatures of 120 degrees in the shade!). I've had close encounters with water cobras, carpet vipers, rattlesnakes, scorpions, and lions. I've worked in grizzly country, which terrifies me—all my fossil photos have bear spray for scale. Then there are the tropical diseases. While working at the Laetoli fossil beds, Tanzania, I got malaria and spent a week in the local hospital. . . . Then, there are the many GI ailments, where you're not really sure what you've got, but given time and cipro, you generally get better. On several occasions, I've made it back to the U.S. with continued GI system failures—on one occasion, I had stopped in D.C. for a meeting at the Smithsonian. I was weak enough that I had to give my talk sitting down.

BOX 4.5

Ellen Currano grew up in the city. "I was introduced to dinosaurs in first grade, and I thought they were the coolest things ever. My parents encouraged my interest in so many ways, including regular visits to the Field Museum in Chicago. Growing up in a huge city, I dreamed of getting to spend time in the wilderness, of going into the field and collecting fossils. At the University of Chicago, I majored in geophysical sciences because of the department's awesome field trips—I got to go to the Bahamas, Big Bend and the Guadalupe Mountains, and Iceland! My junior year, I studied abroad in Tanzania and began a lifelong love affair with East Africa. The summer after my junior year, I was an intern at the Smithsonian under paleobotanist extraordinaire Dr. Scott Wing. We spent a month in the Bighorn Basin, Wyoming, doing fieldwork. I found my calling in the middle of the Wyoming desert excavating beautiful leaf fossils, and I realized that, for me at least, the research that could be done using leaf fossils was far more interesting than anything that could be done using dinosaurs. That summer I met my future PhD advisor at Penn State, Peter Wilf. After college, I went to Penn State for my PhD, coadvised by Scott and Peter. Every summer, I returned to the Bighorn Basin for more fieldwork and savored every minute of it."

And then there are the manmade dangers. The "real Indiana Jones" was the vertebrate paleontologist Roy Chapman Andrews, whose accounts of his exploits in the 1920s have always thrilled me. Among his most famous finds were nests of dinosaur eggs in Mongolia. Andrews is often pictured sitting with a rifle on his lap. This was to protect himself and his crews from bandits who roamed the politically unstable area. Gaylord Simpson found himself in the middle of a revolution in Argentina in 1930 and narrowly avoided getting shot to death. These dangers are not just a romantic sounding vision from the past. In 2008, a field party led by a renowned paleontologist was robbed by bandits in Niger, including a shot fired over their heads.[14] Fieldwork in the fossil rich Siwalik Mountains of Pakistan has essentially stopped due to the unrest there. Closer to home, there are always risks of stray shots from hunters or target practice. The best way to avoid or minimize risks is to anticipate them and, as a colleague of mine tells students, "don't be stupid."

Problems also can come from your companions in the field. Fieldwork in paleontology, as in many other areas of science, has an unfortunate legacy of sexual harassment and even sexual assault. One female colleague angrily recalls being propositioned and grabbed by a well-known vertebrate paleontologist, "one of five" known by woman paleontologists to be serial offenders.

It was one reason she left vertebrate paleontology to become an invertebrate paleontologist. Another recalls: "A close friend and colleague (a graduate student at the time) had shared with me two separate instances where she was sexually harassed by older, more established male colleagues in a field setting. Both instances deeply troubled her, and she felt she had nowhere to turn, and no one to turn to. . . . I think because many of us have a field component to our research, the opportunity for harassment is greater than in other fields." A 2014 survey examined the experiences of women and some men early career scientists ("trainees') in field settings.[15] About half of the 666 respondents were anthropologists; the remainder included geologists, archeologists, and ecologists, some thirty-two disciplines in all. About two-thirds of the respondents described some form of sexual harassment, e.g., inappropriate remarks, jokes or comments, and a shocking 20 percent reported some form of sexual assault, such as unwanted sexual contact. I suspect that the numbers in paleontology, at least historically, are not markedly different.

For many years, beer has been treated as one essential. Although drinking has long been ingrained as part of fieldwork, this has come under increased scrutiny as being detrimental. I have seen and heard about concerning behavior by drunk students and the faculty who are responsible for their safety. This is another area of our culture that needs scrutiny.

There is hope that things are getting better. Part of this is the increasing willingness among victims and witnesses of harassment to speak out and act. Another part is the recognition within the community that this situation exists and can no longer be ignored. Lucy Edwards, a long-time micropaleontologist at the United States Geological Survey, points out that a major change in the status of woman in the field is that "unwanted touching and crude or disparaging comments are now legally recognized as part of a hostile work environment." The major paleontological societies now have explicit statements about sexual harassment and abuse, and the Paleontological Society adopted a rigorous code of conduct in 2019.[16] As Phoebe Cohen said in the Winter 2016 edition of *Priscum*, the Paleontological Society newsletter: "It is important for us to send a clear message that such behavior is unacceptable and to also send the message to those who may have experienced discrimination or harassment that the Society takes such incidents seriously."

It would be a serious misconception to believe that all paleontological fieldwork takes place on land. The oceans cover about 72 percent of the world's surface area, and the vast majority of the rock record lies underneath what can be tens of kilometers of water. Beginning in 1968, a series of programs, using increasingly sophisticated ships and drilling technology, have drilled tens of thousands of cores into the ocean floor, capturing rock records going back tens of millions of years. The data from these cores and those collected by other oceanographic vessels laid the foundation for the acceptance of the theory of plate tectonics. They also provide crucial data for reconstructing the history of the world's oceans and climates, such as the timing and magnitude of the advance and retreats of the glaciers during the ice age. Paleontologists aboard these ships are critical parts of the scientific teams that analyze the cores when they are brought on board and when they are returned for further analysis on land.

Pincelli (Celli) Hull (Yale University; box 4.6) is a micropaleontologist who has worked as a sedimentologist on several deep-sea drilling trips. As she describes it, it is very different from the typical experience on land. Given the expense of running the ships, they operate twenty-four hours a day and are very loud at all times. The life of a scientist on board is extremely regimented, with a twelve hour on/twelve hour off duty cycle, although the actual workday can be fourteen to sixteen hours long. So that they are not distracted, all the daily needs of the scientists are taken care of by the crew. The ship must stay in place to complete drilling in one spot, so it shudders constantly. It is also a highly organized enterprise, with an assembly line approach to drilling the cores, recovering them from the seafloor, bringing them into the lab, describing what is in them, and storing them in preparation for the next core. The job of the micropaleontologists is to date the sediments in the core while the sedimentologists describe their physical properties. The results are written up and prepared for publication by the time the ship docks, some two months after leaving home. Despite the hard work and long hours, Celli is quite enthusiastic about it: "It's all about discovery!"

BOX 4.6

Pincelli Hull is a bit of a latecomer to paleontology. As a college student at Duke, she was interested in oceanography, double majoring in biology and earth and ocean sciences. She spent her summers at Duke's marine lab. She went on to graduate school at the Scripps Institute of Oceanography in biological oceanography, but she realized that ecological time series were not long enough to study some of the more interesting patterns in the evolution of the single-celled organisms she was studying, so she turned to the fossil record and the study of mass extinctions.

Other paleontologists do work much closer to shore at various marine laboratories. In these cases, much of our work overlaps with that of our friends and colleagues in marine biology. Examining living forms and their habitats is essential training for understanding the same relationships in ancient seas, and I require it of most of my graduate students.

Arnold Miller (University of Cincinnati) has extensively scuba dived the area around St. Croix in the U.S. Virgin Islands with the goal of understanding what controls the distribution of shells and other skeletal remains on the seafloor, especially the role of storms. He ended up getting firsthand experience with this when category 4 Hurricane Hugo struck while he was there in 1989, severely damaging the West Indies Laboratory for Underwater Research where he was working. Arnie recalls that several times the roof of the building where they sheltered seemed certain to fly off.[17]

Fieldwork in modern environments also takes place far from the ocean. Kay Behrensmeyer (box 4.7) has conducted extensive vertebrate paleontological fieldwork in the Siwalik mountains of Pakistan. She has also worked closely with paleoanthropologists on the paleoecology of early human sites in Africa. But for many of us, she is best known for her long-term study of modern vertebrate remains in Amboseli Park, Kenya, focusing on the processes involved in fossilization. Since 1975, working with the ecologist David Western, she has periodically visited the park, tracking the breakdown of carcasses and bones on the land surface.[18] Her work has been key to understanding how vertebrate remains become recycled or preserved in terrestrial settings. Kay's research is one of the pioneering studies in modern approaches to taphonomy, the science of fossil preservation.

BOX 4.7

Kay Behrensmeyer, curator at the National Museum of Natural History (NMNH), has been an inspiration and mentor to many younger scientists. She recalls that "from an early age, I was interested in fossils and rocks, and natural history in general. My two brothers and I collected Paleozoic marine fossils where I grew up in western Illinois, along a creek on the farm where we spent our weekends. My parents and my aunts encouraged us to be good observers and to be interested in science. They gave us books and specimens of rocks and fossils, and we had National Geographic Magazine *to inspire us about the world beyond the Midwestern USA. My father had an avid interest in science, and around the dinner table he told us stories about scientific discoveries. When I went to college, I intended to study art, but a really great geology course made me realize that I could follow my interest in fossils and rocks to discover new things about the history of life. I also had a few dreams about being part of fossil-hunting expeditions to exciting places around the world. Then I went to the Indiana University Geological Field Camp in Montana, learned to map and work out geological history, and my career trajectory was on its way." Kay went on to earn her PhD from Harvard.*

Inspired by Behrensmeyer's approach, University of Georgia paleontologist Sally Walker and colleagues have conducted long-term experiments on the fate of shells on the seafloor.[19] In 1993 they placed mollusk and brachiopod shells, crabs, urchins, and wood on the seafloor, spanning shelf-to-slope habitats in the Bahamas and in the Gulf of Mexico, and they have recovered them at intervals since that time. Given that the depths ranged to more than 600 meters, a submersible was used to put the experiments in place and to recover them. Sally describes her first experience diving to 610 meters (2,000 feet) in the Gulf of Mexico as one of the highlights of her life: "Imagine climbing into a goldfish bowl, except I was the goldfish surrounded by 3-inch Plexiglas. Safety checks were finalized, communications were established, and the descent below the waves began." Soon Sally "was in the "twilight zone," traveling through a "murky darkness that soon became an ink black night."

The red lights of the instrument panel indicated that the sub was dropping 30 meters (100 feet) a minute. A few minutes later, near the target spot to deploy experiments, the sub's pilot turned on the arc lights and I peered out at the surrounding underwater scene full of almost swimming pool-sized pits pocking the seafloor. "What the heck are those huge pits?" I asked. The pilot calmly replied, "Those are methane blow-out pits." When the underwater gyres heat up during a warm-climate interval, the frozen methane under the sediments turns to gas and blow out these huge pits. We hovered over the seafloor to an amazing array of large sea fans and giant straw-like tube worms surrounded by a wreath of mussels that were near where we would deploy our experiments . . . so much life at 610 meters! We were startled, however, to see an undulating body larger than our sub materialize at the edge of our vision: it soon became clear that it was a seven-gilled shark. We also saw silver-dollar-sized deep-sea hatchet fish, with flat, clear bodies and arrow worms that darted in all directions. We soon completed our task, turned off the arc lights, and started our ascent to the surface. The pilot asked if I wanted to see the 4th of July fireworks. I was confused as the date had passed, but nonetheless I said: "Sure!" He then covered the lit instrument panel with a towel and flashed the arc lights once during our ascent. Tears started to roll from my eyes: the arc lights had stimulated all the little sea creatures to bioluminescence, and there were Carl Sagan billions and billions of biologically produced lights, in all their glory and diversity, shimmering and sparkling at 600 meters in the deep sea! What an amazing adventure!

Although paleontologists have been collecting fossils for centuries, there are always new things to find. Scarcely a month goes by without the public announcement of a new dinosaur, a new ancient mammal, or new sites in the Cambrian. Many of these come from sites in remote areas, but as my cave discovery shows, it is not unlikely to find them in places far closer to home. Stunning new discoveries have come from China, which is in a golden age for paleontology, and from Morocco. It is summer as I write this. Most of my colleagues are in the field, finding new things. I am getting itchy to abandon my keyboard and join them.

5

SAFE PLACES

One of the most famous sites in the world is almost in my backyard. The various sites that comprise Mazon Creek, named for the place where its fossils were first found, preserve a huge number of terrestrial and marine organisms. Many people who grew up in the Chicago region own at least one preserved leaf from the site, and for decades there has been an entire community of avid amateur collectors. One strange organism from Mazon Creek, the Tully monster (*Tullimonstrum gregarium*), was named the state fossil of Illinois long before its identity was established. Two papers published in 2016 came up with the controversial result that it was a vertebrate. My own interest in the site is in the most common animal found there, beasties the collectors call "blobs," which was named *Essexella* in 1979 and identified as a jellyfish. For a host of reasons, I became dubious about the identification and decided to restudy the fossil.

To do this, I did not go out into the field and collect new samples of *Essexella*. The reason was simple; literally hundreds of specimens, most collected by amateurs, were already stored and cataloged at the Field Museum. Included in this group were many of the specimens used by Merrill Foster in 1979 to establish his identification. Studying these specimens was much easier and faster than going out to look for my own specimens at the site. The group that worked on the Tully monster did the same; they relied on the numerous specimens that had been collected, many by amateurs, and placed safely in the museum.

It is impossible to overestimate the importance of the world's natural history museums. They store, preserve, record, and exhibit the diversity of our natural world. Items that are lost from nature today still exist in the cabinets and displays of our museums. On the first floor of the Field Museum, you can gaze on mounted specimens of passenger pigeons, some of the last remnants of a species that once numbered in the billions. The collections of the Chicago Academy of Sciences, a local natural history museum whose roots long predate the Field Museum, contain trilobites and mammoths from sites long paved over or filled in. On display at the Yale Peabody Museum is the very first *Brontosaurus* to be discovered and named, the holotype specimen of *Brontosaurus excelsus* Marsh 1879 (YPM 1980).[1] You cannot discuss *Brontosaurus*, or any of the related dinosaurs, without referring to this specimen. Natural history museums are the direct physical record of the physical, biological, and human world over time and can never be replaced by even the most sophisticated digital technology (although they use it extensively). They are also the first exposure to the wonders of science for many children, especially for those in urban areas, and they remain a source of wonder and beauty for adults.

The public persona of paleontologists is inseparable from museums. The image of the paleontologist carefully putting dinosaur bones (of course) together has been used frequently in comics, television shows, and movies, perhaps most famously by Cary Grant in *Bringing Up Baby* and his manic search for the "intercostal clavicle" needed to finish his *Brontosaurus* skeleton. Real paleontologists at museums are, of course, nothing like the shy and befuddled David Huxley portrayed by Grant (they are, of course, just as attractive).

Major natural history museums, such as the American Museum of Natural History, the National Museum of Natural History, and the Field Museum in Chicago employ multiple paleontologists. The top ranks, roughly equivalent to university professors, are curators. Curators have multiple tasks: they are ultimately responsible for the vast fossil collections in the museum; they supervise the staff in their area of responsibility; they work with the exhibition departments to make sure the museum's exhibits are accurate and up to date; they meet with members of the public to answer questions; and they conduct cutting edge research on the fossil record. Many teach classes at local universities and may even supervise students of their own. And they are often called on to help raise the funds that keep the museum operating.

Anna Kay Behrensmeyer is a curator at the Smithsonian's National Museum of Natural History: "As a museum scientist involved in a lot of outreach and education, I see paleontology as a continuing source of fascination for the public of all ages. Many of our visitors seem to be well-informed these days, with thoughtful questions reflecting basic knowledge about fossils and ancient life." Kay clearly loves her job: "The museum is an amazing place to work because of the NMNH community of scholars, the collections, our continual flow of visiting scientists, and the opportunities to interact with the public. Research curators have the freedom to chart their own research trajectory and are not constrained by an academic schedule, although if we want to teach there are many opportunities."

Reporting to the curators are the collections managers. These critical individuals are responsible for the day-to-day maintenance of the collections. This includes making sure the fossils are safely and securely stored and are properly cataloged. Collections managers provide access to the collections for visiting scientists and arrange for loans of materials for those who cannot directly visit. Their intimate knowledge of the collections is invaluable. I could not have written some of my recent papers without the help of collection managers Paul Mayer (Field Museum), Jen Bauer (University of Michigan Museum of Paleontology), and Peggy Fisherkeller (Indiana State Museum). They also document the use and impact of the collection, strategize for its future development, manage and improve access to collection data for scientists and the public, and acquire new collections. Many also interact directly with the public and can be the public face of the museum. If time and museum policy allow, some collection managers conduct their own research. Paul Mayer was a coauthor on the Tully monster paper and actively interacts with the local fossil collecting community and the public.[2]

Susan Butts is the invertebrate paleontology collections manager at the Yale Peabody Museum; her official title is senior collections manager (figure 5.1). She characterizes herself as an "accidental paleontologist." She grew up in Connecticut, a state not known for its fossils. Susan went to college as a

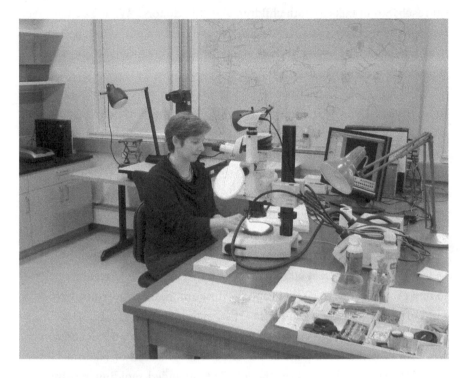

FIGURE 5.1 Susan Butts photographing a specimen under the microscope.
Source: Photo by the author.

political science major to become a diplomat. She took Geology 101 to meet
her science requirement and decided that "geology was awesome." She immediately changed her major. Although her paleontology class was lab based and
didn't convert her immediately, Susan got excited by brachiopods she saw in
a waterfall in upstate New York where she was studying stream chemistry for
an independent study. She eventually did a PhD at the University of Idaho,
before being brought in as a postdoc at Yale. Susan somewhat reluctantly
accepted the job as the collection's manager, a job she now loves. She also says
she would not do it anywhere but at Yale.

Butts calls being a collections manager the "weirdest job ever." In addition
to the tasks already described, she must be a hostess to visitors, a spokesper-
son for the museum, and perform maintenance (including cleaning out dust
bunnies) in addition to being a scientist. Some of what she does is public

outreach; Susan talks to both children and "big donors." What Yale has given her is autonomy; she can be a principal investigator on grants and do her own research on brachiopods, although she never has time. In many other institutions, she would be a curator.

Although many collections managers have master's degrees, more and more have PhDs in paleontology. They also have more responsibilities. They are active in international organizations and serve on National Science Foundation panels. Many belong to the Society for the Preservation of Natural History Collections[3] where they share best practices related to the management and conservation of natural history collections, including digitization.

Curators and collections managers rarely do the exacting and slow work of removing fossils from the rock and assembling the broken pieces This and related work are the job of preparators, who are highly trained and skilled as well as amazingly patient and focused. Marilyn Fox is the vertebrate paleontology preparator at the Yale Peabody Museum. Marilyn was not trained as a scientist; instead, she began as an artist, making prints, etchings, and lithographs and earning her master's of fine arts. She ended up working on a major renovation of the fossil halls at the American Museum of Natural History (AMNH), and she learned on the job. This included molding, casting, and stripping old varnish from the renowned *Tyrannosaurus rex* at the AMNH. Marilyn thinks of fossils as data; in her words, "preparators expose and preserve the information in the fossil."

The work of a preparator can be highly technical. They use a wide variety of tools and techniques and must have knowledge of basic physics, chemistry, biology, anatomy, and chemistry, as well as material science. They are both scientists and artists. But as Marilyn Fox puts it, "critical thinking is the essential competence for vertebrate paleontology preparators." When preparators meet, Marilyn notes that much of their discussion is about new high-tech preparation methods, and professional paleontologists are usually unaware of these methods. She is active in the Association for Materials and Methods in Paleontology, a small professional society that shares ideas, standards, and best ethical practices.[4] By the way, one thing that particularly irks Marilyn is the use of the word "dusty" to describe museum collections. In any well-maintained museum, the specimens, drawers, and cabinets and the spaces they are in are anything but "dusty." Lose the cliché!

In addition to the large natural history museums, many smaller museums also employ paleontologists. These may be state museums, such as the New

York State Museum, museums attached to major universities, or even stand-alone museums. Examples of these are the Burpee Museum in Rockford, Illinois, and the Dinosaur Discovery Site in St. Georges, Utah. These museums may have only one or two paleontologists and a few staff. Patricia (PJ) Burke has the joint title curator of geology collections/senior collection manager at the Milwaukee Public Museum, where she is the sole staff member in her department. "I grew up along the Mississippi River tromping about looking for arrowheads, tadpoles, clam shells, and generally cool stuff. I thought I would be a biologist, but I tripped into a plant evolution class in college and was hooked. I took a few geology classes and figured out that I could combine biology and geology in paleontology. Museums are the fun part of science for many people. . . . We have the real stuff. The collections can be both an opportunity and a challenge. The care and study of collections is a financial burden, but they also fuel the knowledge behind the exhibits and outreach."

In addition to the professional staff, many museums are fortunate to a have a cadre of knowledgeable and enthusiastic volunteers. Some of these are skilled preparators, such as Kurt Zahnle at the Field Museum, others lead tours or help take photos of specimens. They do not get as much credit as they should.

One of the fun things to do when in the collections area of any museum is to randomly pull open a drawer and see what is in it. Many years ago, I did just that with a specimen drawer at the Field Museum and saw the handwritten note "brought over by Lyell." I was stunned. Charles Lyell (1797–1875) was one of the founders of modern geology and one of the most influential figures in the history of science. His writings were seminal for the work of Darwin, who brought a copy of Lyell's *Principles of Geology* with him on the *Beagle*. In 1841 and 1845, Lyell visited New York, where he met, went into the field, and argued about geology with James Hall, considered the father of American invertebrate paleontology. Much of Hall's fossil collections ended up at the Field Museum, and the specimens I found had been brought over by Lyell to show Hall. For a paleontologist, these are the closest thing to holy relics.

The point here is that you never know what treasures a museum collection might contain and what answers they may provide. As Jessica Theodor (University of Calgary) asserts, it is important to realize that many paleontologists work with fossils that have already been collected. For example, the tiny fossils known as conodonts were for decades one of the foremost mysteries in paleontology. As the name implies, they come in a variety of forms that resemble teeth. Very rarely, they preserve as complex bilaterally symmetrical assemblages of multiple conodonts. Quite abundant in rocks from the Cambrian to the Triassic, when they disappear, they have long been useful in dating rocks (biostratigraphy) because they evolved rapidly and are easily identifiable. But we did not know what kind of animal possessed the conodonts, except it must be small, bilateral, and probably free-living. Several possible "conodont animals" were suggested over the years, only to be rejected based on further investigation. The break came in the early 1980s when Derek Briggs (box 5.1) and Euan Clarkson were looking through the collections at the Institute of Geological Sciences in Edinburgh, Scotland, for specimens of shrimp and other animals from a Carboniferous formation in Scotland. In their search, they came across the fossil of a wormlike animal that had probably been collected in the 1920s. Careful work on the minute specimen eventually revealed that it was the long-sought conodont animal and a member of our own group, the vertebrates. Since then, about a dozen additional specimens have been discovered worldwide. Mystery solved—thanks to a museum collection.

BOX 5.1

---

*Despite his stellar reputation and many awards in the field, **Derek Briggs** (Yale University) did not start out as a paleontologist. He was raised "in granite," the bedrock where he was born in Ireland. Interested in natural history, he attended Trinity College in Dublin, where he "read science" (undertook a course of study). Derek decided to major in geology because "geologists were the most interesting people," and he found that paleontology was the most interesting class. He left Ireland for the broader opportunities offered by British universities and ended up at Cambridge because he had to go where private funding was available (he was not eligible for government funding). His advisor at Cambridge was the great Harry Whittington who, along with Derek and his fellow student Simon Conway Morris, transformed our knowledge of the Cambrian fossil locality known as the Burgess Shale. Their work and interactions were eloquently described by Steve Gould in his best-known book,* Wonderful Life.

As in the rest of science and society, technology has radically transformed how museums conduct their work. Some of these changes are fairly simple; the introduction of high-density cabinet systems has greatly increased the storage capacity of collections without requiring additional floor space. Paper museum catalogs, often written in exquisite penmanship, have been converted to online and searchable form. This has greatly eased the problem of locating which museum has which specimen and where it is stored. An exciting recent development goes under the name iDigBio, which stands for Integrated Digitized Biocollections. Funded by the National Science Foundation, iDigBio is a multiyear, multi-institution effort to produce a single electronic resource for the tens of millions of biological specimens, including fossils, housed in the nation's natural history museums and herbaria. An associated project is iDigPaleo, which is focusing on collections of fossil insects.

The amount of work needed to produce the iDigBio database is amazing. As of July 2019, information on more 100 million specimens had been entered. About 20 percent of these have some sort of accompanying image. The specimen records also include the critical data on where and when the specimens were collected. All of these can be searched through a well-designed portal. For a scientist planning a research project, the ability to know what is in museum collections prior to a visit or to request a loan of materials is an invaluable aid.

Museums have two main complementary, although occasionally conflicting, missions. The first of these is to be a critical part of the infrastructure of science, as important as any large telescope, particle accelerator, or research hospital. Fundamental research on our natural world, such as understanding current and past biodiversity and thus potentially the future, is carried out by scientists employed by or using our museum facilities. The second mission is to educate the public through exhibits, classes, and direct contact or outreach.

If you see photos of museums from early in the last century, you will see rows upon rows of labeled specimens in glass cases. These would often be supplemented by exquisite dioramas or beautiful artwork showing a vision of past life. In the latter part of the twentieth century, however, it was decided that this style of museum was simply not interesting to the public, who were used to the instant entertainment provided by television and the developing internet. As a result, many of the old exhibits were torn out and replaced with more "interactive" and "entertaining" displays. In many cases, the curatorial staff were shut out of the planning of the exhibits. At the Field Museum, the longtime fossil hall was replaced by a truly atrocious new exhibit. Children could sit on little moveable chairs that looked like trilobites, but there were no actual trilobites on display. The idea that museums should put at least part of their collections on view seemed to be missing.[5]

I am thankful that this trend seems to have slowed or even been reversed. The horrible exhibit at the Field Museum has been replaced by a far better new one that emphasizes the science and displays the fossils. The National Museum of Natural History in Washington, D.C. opened a new Deep Time Fossil Hall in 2019, after a five-year, $48 million renovation. Preparing the new hall consumed much of the time and effort of curator Kay Behrensmeyer and her colleagues. The Yale Peabody Museum is undergoing a major expansion and renovation, with close involvement by Jacques Gauthier, Susan Butts, and other members of the scientific staff.

For paleontological exhibits, it is especially important to impart a basic understanding of evolution, as well as climate change. Jacques Gauthier (curator at the Yale Peabody Museum; box 5.2) rightly insists that evolution should be at the forefront of every paleontological exhibit. Bruce MacFadden, a curator at the Florida Museum of Natural History, agrees with Gauthier about the importance for teaching evolution. He blames the lack of public understanding of evolutionary concepts on the scientists and museums themselves, who have not been good communicators. He is especially concerned that the evolution of groups has been presented as straight-line progressions rather than the bush they actually are. This is something he is intimately familiar with because he is an expert on the bushy evolution of horses.

BOX 5.2

Jacques Gauthier, a vertebrate paleontologist and herpetologist at Yale University, is a product of the paleontology program at the University of California, Berkeley. His formative moment as a child was seeing Disney's Fantasia, *in particular the "Rite of Spring" section, which includes dinosaurs and their extinction. Jacques remembers his dad explaining to him that what he was seeing were dinosaurs. Jacques grew up with wildlife just outside his door in the Mojave*

Desert of California, chasing lizards by day and snakes at night, something he still does as part of his research. He went to San Diego State University to get both his bachelor's and master's degrees because they had the leading experts on lizards. It was while doing his master's that he saw some Eocene fossil lizards and first became interested in the fossil record. Going to UC Berkeley in paleontology, he admits, required a "learning curve on deep time." The curve was conquered, and he is now one of the world's most renowned workers on dinosaurs, lizards, and their living and fossil relatives.

I have known Bruce since my undergraduate days at Columbia and the American Museum of Natural History when he was earning his PhD under Malcolm McKenna. He is the only scientist I know who has been president of both the Society for Vertebrate Paleontology and the Paleontological Society. Like many others, he had a childhood interest in dinosaurs and fossils, which was supported by his single mom who took him on the train to the AMNH. He drifted away from this initial interest, however, and ended up going to chef school. After two years at the University of Maryland, Bruce transferred to Cornell University, where he was greatly influenced by paleontologist John Wells, who gave him the confidence to go on to graduate school. Bruce has spent virtually all of his career at the Florida Museum of Natural History and has become particularly active in outreach, including founding The FOSSIL Project, which uses social media to increase interactions among amateur and professional paleontologists. (I return to this later.)

The University of California Museum of Paleontology (UCMP) considers itself to be a research museum and has very limited public displays. What it does have, however, is the superb Understanding Evolution and Understanding Science websites on paleontology, evolution, and teaching science. I send teachers and students to these websites when they are looking for reliable information on these topics. The websites were originally developed by the wonderful Judy Scotchmoor, then assistant director of education and public programs.

In 2013, Judy retired and was succeeded by the equally amazing Lisa White. Lisa is a micropaleontologist, director of education and outreach at UCMP, and one of the tiny number of Black paleontologists. She has long been dedicated to increasing diversity in the geosciences. A native California, Lisa grew up in San Francisco, close to the California Academy of Sciences, which she credits with sparking her early interest in geology and paleontology. Inspired by Ansell Adams, Lisa began her academic career as an arts major and planned to go into photography. To meet a general education requirement at San Francisco State University, she took her first geology class with "a really terrific instructor." This led to an internship at the United States Geological Survey Office in Menlo Park, and she went on to earn her PhD at the University of California, Santa Cruz. Before coming to UCMP, Lisa was a faculty member at San Francisco State University for twenty-two years, where she directed SF-ROCKS (Reaching Out to Communities and Kids with Science in San Francisco) and SF-METALS (Minority Education through Teaching and Learning in the Sciences) geoscience programs. She is especially concerned that we need innovative ways to apply our training to excite people about science. Joining UCMP has given Lisa an opportunity to interact with a diverse research community, allowing the incorporation of "authentic research in an array of UCMP learning modules and materials." She recently helped develop a website at the University of California, Berkeley on Understanding Global Change, an issue she feels will help attract minorities to science.[6] The UCMP is inspiring a new generation of scientists to focus on science communication (box 5.3).

BOX 5.3

Sara ElShafie is a PhD student at the University of California, Berkeley, working with renowned vertebrate paleontologist Kevin Padian. When Sara enrolled as an undergraduate at the University of Chicago, she knew nothing about paleontology. By chance, she ended up volunteering for and then being hired for three years by famed dinosaur hunter Paul Sereno to "clean dinosaur bones." While working with Sereno's Project Exploration, Sara also saw firsthand how paleontology excited kids. Sara went on to get a master's degree at the University of Nebraska–Lincoln and then enrolled at UC Berkeley. She describes herself as "an evolutionary biologist with interests in vertebrate paleontology, herpetology, global change, and conservation biology," with a current focus on lizards. Sara also runs a personal business on science storytelling, inspired by the film industry. Her long-term goal is to be a museum director and be involved in science education and communication.

Sara's inspiration for this is the paleobotanist Kirk Johnson, who is director of the Smithsonian National Museum of Natural History. Kirk is an outstanding scientist, and he also has to manage a large enterprise of some 450 employees. In addition, he has appeared and hosted multiple times on Nova, the PBS documentary series, and has written some delightful books with Ray Troll, a highly creative paleoartist. According to Sara, "[Kirk] encouraged me to get a PhD in science, as well as outstanding outreach and communication experience (informal education), and to take a business course."

Another institution with a strong public outreach program is the Museum of the Earth of the Paleontological Research Institution (PRI), in Ithaca, New York. When I first joined the PRI in the late 1970s, it was a sleepy, research-oriented entity housed in a former mansion. Its main mission was to house its extensive fossil collection and to publish an important series of highly technical publications on fossil groups. This all changed when Warren Allmon became director in 1992. Under Allmon's leadership, PRI raised funds for and constructed a beautiful new public museum that opened in 2003. It has also begun a broad range of educational and outreach activities, including major "citizen science" programs. PRI also has a large permanent education staff, who have been highly influential in development of the *Next Generation Science Standards*, a widely adapted framework for teaching science in the nation's public schools. All this has been accomplished while maintaining PRI's scientific work. No longer sleepy!

Patricia Burke at the Milwaukee Public Museum points out that "we still seem to be the place to get questions answered and things identified. I don't think there has been a week in my thirty-year career when I have not had a public inquiry. MPM's visitor demographics show an amazing diversity. We can reach families, elderly, young, and multiethnic populations."

The Burpee Museum in Rockford, Illinois, is a long-time leader in public outreach. Much of the credit for this goes to Scott Williams, who in 2003 took over their annual PaleoFest, changing the format and increasing its visibility. The tremendously successful PaleoFests are two-day public educational events that feature numerous guest speakers and exhibits, with a strong focus on dinosaurs. Scott is now at the Museum of the Rockies.

Scott grew up near Rockford in the small town of Stillman Valley, Illinois, where he developed an early interest in geology and paleontology. His parents took him to the Field Museum, the Burpee Museum, and on a trip to visit museums and national parks in the West. Starting at the age of thirteen and continuing for seven years, he volunteered at the Burpee. After graduating from the local community college, Scott pursued a career in law enforcement but still volunteered on weekends at the Burpee. Along with curator Michael Henderson, he organized, funded, and planned field expeditions to Montana, which resulted in the discovery and collection of "Jane," the Burpee Museum's *T. rex*. He eventually left the sheriff's office to become the Burpee's fossil preparation lab manager and eventually director of science and exhibits. In addition to his work on PaleoFest, Scott began National Fossil Day in cooperation with the National Park Service. National Fossil Day is "an annual [October] celebration held to highlight the scientific and educational value of paleontology and the importance of preserving fossils for future generations." It involves numerous national parks, museums, fossil clubs, and other organizations.

The Burpee Museum is a good example of the many small to medium-sized museums that dot the country and employ paleontologists. Although they may not be as well-known as the American Museum of Natural History or the Field Museum, they often have significant fossil collections, an active research program, and strong public outreach. Examples include the Museum of the Rockies (Bozeman, Montana), the Raymond M. Alf Museum of Paleontology (Claremont, California), the Mammoth Site (Hot Springs. South Dakota), and the St. George Dinosaur Discovery Site (St. George, Utah).

I recently had the opportunity to be a part of "ID Day" at the Yale Peabody Museum, a day when members of the public bring their items to the museum and experts in various fields identify them. I was at the Invertebrate Paleontology table with Susan Butts. In preparation, Susan had brought specimens from the Devonian of New York. I, on the other hand, brought a tray of pseudofossils. What are pseudofossils? They are naturally occurring inorganic structures in rocks that fortuitously resemble living things and thus can be mistakenly identified as fossils. For example, dendrites are branching patterns produced by deposition of minerals, usually a manganese oxide, along the bedding planes of sedimentary rocks. They can easily be confused with some sort of plant. Other pseudofossils resemble bones, teeth, eggs, turtle shells, or whatever the imagination provides.[7]

Why did I bring pseudofossils? Anyone who has worked at a museum or in a geology department is familiar with the phenomenon of a visitor (or, more recently, an email correspondent) who is convinced that an amazing find has been discovered that you, as an expert, need to confirm. For geologists, this is often a piece of rock that the visitor is convinced is a meteorite. In all but the very rarest cases, it is not a meteorite but what is tongue-in-cheek called a "meteor-wrong." For paleontologists, this is frequently a concretion identified by an eager member of the public as a dinosaur egg, a dinosaur brain, or a turtle shell.

Sally Shelton is a highly experienced natural history museum collections manager who was at the Museum of Geology and Paleontology of the South Dakota School of Mines and Technology and is now at Texas Tech. She told me that "EVERYTHING seems to be fossilized brains. Our rock garden, where the treasures go to die, could probably qualify for Mensa." So I brought the pseudofossils in anticipation that someone might bring such a treasure. At the least, I could teach what pseudofossils were.

Most eager collectors of pseudofossils, albeit disappointed that their rare and valuable artifact is a chunk of chert or a stream-weathered piece of limestone, take this news with good humor. And we emphasize to them the inherent scientific interest that even such a seemingly mundane object has. Some collectors are harder to convince. I had a seemingly endless series of emails with someone who was convinced his limestone concretion was a turtle skull with a preserved eye. Another took some convincing that his bed of chert nodules was not a dinosaur graveyard. But there is a small group who become combative when the expert does not tell them what they want to hear. Again, Sally Shelton explains: "the hostile crazies with nonfossils in cement chunks, meteor-wrongs, gemstones looking suspiciously like Coke bottle shards, and other items that they believe to be priceless treasures. . . . We can handle diplomatic identifications, but the hostility is getting old. . . . Clearly, we just want to get our hands on their treasures and keep them for ourselves, so we tell them that their treasures are not what they know they actually are. Because that's what museum people do." Vertebrate paleontologist Denny Diveley was at the Idaho Museum of Natural History when "a red 'petrified' chicken heart walked into the IMNH . . . the proud owner would not accept the gentle suggestion 'chert.' My bad." Rod Pellegrini (New Jersey State Museum) recalls "a guy that dropped by on us a few years ago with a quartz pebble he swore was a crystallized dinosaur skull that had a clear tooth with blood remnants and clear tooth decay. . . . But he did learn one thing: it was made of quartz, which after much online 'research' he found had special properties, including energy-related powers and vibrational qualities. We had been clearly lying to him because the tooth cavity and dried blood were plain as day, and anyone could see that."

These are fun stories to tell among ourselves. But in these cases, as with all dealings with the public, we are unfailingly polite and helpful. There is always the rare chance the perceived treasure actually is one and something that deserves scientific study. And it does not serve us, science, or the public to be snarky.

When asked about challenges facing natural history museums, Patricia (PJ) Burke states: "I have to start with finances since it seems just about every U.S. natural history museum is facing some type of funding challenge. Large, small, nonprofit, government, university, all are facing financial issues. Museums seem to be facing a life-or-death challenge." PJ has direct knowledge of this challenge; the Milwaukee Public Museum nearly went bankrupt due to mismanagement in 2005. A $7 million budget deficit was projected for fiscal 2005, leading to a layoff of fifty-six employees, including nearly all of the curatorial staff. Geology collections manager Paul Mayer eventually went to the Field Museum, where he is the fossil invertebrate collections manager. The paleontology curator, Peter Sheehan, was supported by a separate endowment and stayed until he retired. Despite improvements, problems remain; another round of layoffs occurred in 2017.

The Field Museum has had its own set of problems. In 2002 it issued a $90 million bond to pay for new exhibits, a new entrance, and a state-of-the-art collection center. This doubled its bond debt and was part of $254 million in capital spending between 2001 and 2011. This spending and the bond issue were based on optimistic assumptions about investment returns, fund-raising, and increased attendance—and then there was the Great Recession. Unfortunately, by 2013 the museum was about $170 million in debt. The result was a large cut in the operations budget for science that badly damaged the scientific mission of the museum. Some curators retired early; open positions were not filled; morale plummeted. I saw this firsthand. I was on a search committee to hire a replacement for a paleontology curator who had left. We interviewed five candidates and were about to make an offer when the search was suspended. That slot and a number of others have still not been filled. In 2001 the Field Museum had thirty-nine curators; the current number is nineteen. According to Scott Lidgard, an invertebrate paleontology curator, things are getting better, albeit slowly (four new curators are being hired and should all arrive by the end of 2020).

Smaller museums also have had difficulties. At one time, many colleges and universities considered a natural history museum as integral to their educational mission. Many of these have closed over the years. The collections at the Walker Museum of the University of Chicago were absorbed into the Field Museum in 1965. In 1980, Princeton gave its vertebrate paleontology collection to Yale. More recently, the University of Wyoming announced in 2009 that it was closing its geology museum as part of a $18.3 million budget cuts at the university level (the museum had an $80,000/year operating budget). Longtime curator paleontologist Brent Breithaupt was dismissed (he ended up at the Wyoming Bureau of Land Management). This led to an outcry across the paleontological community and the public. The museum was reopened part time, and only with a security guard to keep an eye on the exhibits. A fund-raising campaign led to the reopening of a renovated museum in 2013, but without any full-time staff. In my own state of Illinois, ongoing budget woes led to the temporary shutdown of the Illinois State Museum (ISM) in October 2015. When it reopened in July 2016, it had lost half of its staff, including mastodon and mammoth specialist Chris Widga, who ended up at East Tennessee State University. There is no longer a paleontologist at the ISM, despite its world-class collections.

The threat is not confined to geology and paleontology museums—or even to museums generally. According to a 2015 article in *Nature*, 100 of the 700 herbaria (collections of preserved plant specimens) in the United States have closed since 1997.[8] There is also a decline in the number of specialists on various groups. The number of curators at the National Museum of Natural History has declined by a third since 1993. Universities have also lost their taxonomic specialists, replacing them with researchers in areas such as cell and molecular biology, who are assumed to be able to bring in much greater funding. The result has been a loss of those who are considered to be "old fashioned" natural historians and the decline in natural history courses, such as entomology, ichthyology, invertebrate zoology, or botany. In a detailed 2014 review of the status and importance of natural history, Tewksbury and others discovered that the minimum number of natural history courses required to get a biology degree in most U.S. universities and colleges is zero (yes, zero!).[9] In my institution, my paleontology class is the only place biology majors can get hands-on experience with the diversity of animal life on Earth. We have no herpetologists, ichthyologists, or entomologists. In a time when we are rapidly losing biodiversity, we are also losing those who can describe what we have left. We may never know what we have lost, even though their undescribed remains may sit forever in a museum somewhere. And this will damage us and society. In a must-read 2014 blog post, Jennifer Frazer said that "natural history is dying, and we are all the losers."[10]

The coronavirus epidemic of 2020 has been a disaster for museums. As I write this in late 2020, many natural history museums have been closed for months. Several have furloughed or laid off staff, and the remaining personnel have had their salaries cut. Even a partial reopening will not come close to meeting expenses. The long-term impact remains to be seen.

6

COOL TOYS

When I started at the University of Illinois at Chicago (UIC) in 1982, I was given the "paleontology lab" as my own. State of the art when it was built fifteen years earlier, it was filled with all sorts of equipment that I would almost never use. At the far end was a glove box with a device that sprayed very fine powder at high velocity. This air abrasive unit was designed to gently remove particles of sediment (matrix) from around a fossil. A dental drill sat in one corner, with an attachment that vibrated a sharp needle back and forth at high speed. This again was designed for removing matrix, but somewhat more harshly than did the air abrasive. A rotating motor on a stand was another tool for the same purpose. I was informed that a nasty looking device in one of the drawers was an autopsy saw. There was an ample supply of sharp picks and needles. Much of the equipment was in some state of disrepair and clearly had not been used for years. And the pièce de résistance was this huge and extremely noisy vacuum that was attached to large snakelike tubes that could be bent to suck up whatever dust you generated.

I was thankful that there were some high-quality microscopes, including some fifteen-year-old Wild M-5 binocular scopes. In addition to a camera lucida attachment that aided my poor drawing skills, a special tube allowed one to take pictures with a Polaroid instant camera. With my paltry start-up funds, I added a fiber optics illuminator that produced a bright light where it was needed.

When you enter my lab now, all that is left of that original equipment are the microscopes and the illuminator. Where the air abrasive unit was sits a fish tank with some small snails. Sitting on a counter is a device for measuring forces, with a computer output. Along one wall is a large plexiglas flow tank, which is designed to produce constant flows of water. Inside of it is another electronic device to measure forces. There are several computers and a copy stand for my digital camera. And sitting on a table near where the vacuum was is my Leica microscope setup, which is fully digital and computerized and takes amazing digital images, including 3-D images (figure 6.1). I love this machine. In the room next door are a 3-D scanner and a 3-D printer.

What happened to my laboratory? First, a vertebrate paleontologist designed the lab, with the assumption that it would be used to process samples returned from the field. Although I do some of this, this is not my area of interest. Second, I added equipment to allow me to do the experiments

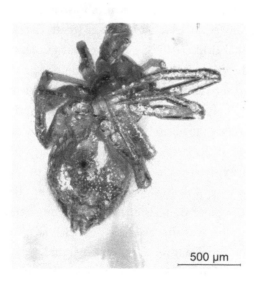

500 µm

FIGURE 6.1 Fossil spider preserved in amber.

Source: Photomicrograph by the author.

in which I *was* interested. Third, some of the equipment was obsolete. Not only was taking a picture with a Polaroid expensive, with no guarantee that it would come out, but they stopped making the film. Most important, as in so many other areas of science and human endeavor, the advent of fast and cheap computing power made much of it obsolete. Paleontology, which might strike you as the quintessential low-tech science, has gone hi-tech. It's not that we don't use hammers, chisels, and fine picks when needed, but they are now supplemented with the most sophisticated apparatus available in science—up to and including particle accelerators. Accompanying this have been welcome interactions with other communities of science and a leap in the sophistication of the questions we can ask about the history of life on Earth.

To understand the influence these modern technologies have had, it is important to understand why they have made such an impact. Every field of science has its classic collection of methods that students are trained to use and that comprise much of the day-to-day activities in the discipline. The 1965 *Handbook of Paleontological Techniques* described such methods for illustrating fossils, especially using photographs. For those of you who grew up with digital cameras, it may be hard to understand how different it was to have to use film. There were tricky technical issues, such as type of film, f-stop, focus, and exposure length. One would not know until the film was developed whether the pictures came out and showed what you wanted. Even if you developed your own film and made you own prints, it was an expensive process from start to finish. And there was no Photoshop to help you "fix" or at least adjust what you shot. Although I miss the magic of seeing a print appear in the developing tray or in a Polaroid picture, I am otherwise not at all nostalgic for film photography. In all the ways that matter to me as a scientist, digital is better. The ability to shoot numerous photos with little cost and to interactively adjust and edit them has been a godsend.

Much of the new technology involves what is called "virtual paleontology," in which fossils are studied through three-dimensional digital images.[1] This is divided into techniques that look at the outer surface of the fossil and those that look at the internal structure (tomographic methods). I recently had a chance to play with an innovative surface technology called reflection transformation imaging (RTI), originally developed for working with archeological and artistic objects. This equipment consists of a dome-shaped array of

FIGURE 6.2 (*left*) Jessica Utrup with the reflection transformation imaging (RTI) array at Yale; (*middle*) Mazon Creek fossil sea anemone imaged with normal digital photography; (*right*) Same specimen imaged with the RTI array.

Source: Photos by the author and Jessica Utrup.

forty-five halogen flashes, with a high-end digital camera at the apex. The fossil is centered at the bottom of the array, and each flash fires in turn, with the images captured by the camera. The software then combines the photos to produce a merged image showing fine surface detail, and I can interactively choose which combination of camera angles produces the best view of the surface. The result is far superior to what can be achieved with a single or even with multiple light sources. I am currently working with fossils from Mazon Creek, which have very little relief. With RTI, I was able to see details I could not see before (figure 6.2).

A related method to RTI is photogrammetry, in which the camera itself is moved. Calibrated photos taken from different locations can be combined for 3-D imaging and measurement.[2] One advantage of photogrammetry is that it can be done with an off-the-shelf digital camera and free software. In addition, it can be used for single specimens or on an entire field site to document the distribution of fossil remains. For example, the positions of the thousands of bones collected and exposed at the Carnegie Quarry at Dinosaur National Monument have been digitized based on photogrammetry.[3] Photogrammetry has also been used to document and study dinosaur tracksites and individual footprints, which cannot be collected and returned to the lab.

Laser and structured light scanning are other technologies that can be used to produce high-resolution images of a surface or even of an entire three-dimensional object.[4] In laser scanning, a focused laser beam is moved over the object, and its reflection is used to map its surface. Structured light scanning projects a series of grids over the object that are captured by a video camera; the pattern of distortions of the grid yields the surface topography. In either case, rotating the object or moving the scanner can capture the entire 3-D form. My own lab uses a structured light scanner, which was far more economical than a laser scanner and proved, after a short learning curve, to be easy to use. The captured image can be converted into a format suitable for a 3-D printer. As a result, I can easily make copies of fragile fossils for use in research or teaching. My colleagues can also share the fossils they have scanned, such as the teeth of the giant extinct Megalodon shark[5] or the bones of the recently discovered hominid *Homo naledi*. This is incredibly useful for teaching and research.[6]

Many traditional techniques involve removing collected fossils from the surrounding rock or sediment without damage. Usually called "fossil preparation," this involves careful mechanical or chemical removal of the surrounding matrix, which is what my lab was originally designed to do. This approach has many shortcomings; it is slow, often requires highly trained professional preparators, can be dangerous, and risks damaging the specimen. Chemical preparation, for example, can use hydrochloric or hydrofluoric acid. The latter is quite hazardous and requires special handling. In many cases the matrix cannot be removed because it provides support for the specimen, which would otherwise disintegrate. The specimen may be integral to the surrounding rock, such as an imprint on the surface. The matrix also often contains important information on the environment or preservation of the fossil. Many of these standard methods were summarized in the *Handbook of Paleontological Techniques*. Over the past fifty years, especially in the last decade or so, new methods have been developed for separating fossils from the matrix that don't involve physically or chemically preparing the specimen. These methods, such as the use of X-rays and electron microscopes, were in their infancy when the handbook was published.

Many of the interesting details of a fossil are well below the surface. A method that has one foot in classic methods and one in modern techniques is serial grinding. In serial grinding, a very thin slice is ground off the rock, and an image of the enclosed fossil is taken. Prior to the advent of digital photographs, this was often a photograph, or a sheet of acetate applied to a surface flooded with acetone—an acetate peel. An additional layer would then be ground off, another image taken, and so on. These could then be combined by hand to produce a 3-D image. Current technology can combine digital images of the ground surfaces to produce a full 3-D rendering of the object that can then be manipulated and studied. The major downside of this method is that it ends up destroying the specimen. For some fossils, however, this is the only practical method to get information.[7]

Ideally, one would like to look below the surface without damaging the fossil. X-rays have been used for this purpose for a long time; there are spectacular x-ray photos of Devonian fossils from Germany and Ordovician trilobites from New York that were made in the 1960s. These were rare, however, and were made possible by the unique preservation of these fossils. They had been turned into pyrite (or "fool's gold"), which provided high contrast in the x-ray images. A two-dimensional image captured most of the information. The method was much less successful with large three-dimensional objects, such as skulls. The major technological advances in this case came out of the medical field, in particular the advent of x-ray computed tomography (CT) also known as CT or CAT scanning.[8] For large objects, standard medical CT scanners can be used; for smaller objects, high-resolution devices known as micro-CT scanners are available.

CT scanning works by taking many X-rays at multiple angles through the object. These are then combined using sophisticated computer software to produce a three-dimensional image. It is also possible to produce a series of virtual "slices" through the specimen, roughly corresponding to the layers exposed by serial grinding but without the destruction of the specimen. The images are wonderfully detailed three-dimensional images of the fossil, allowing detailed reconstruction of the internal anatomy. Like surface light scanning, the image can be used to produce a printed 3-D model. It can also be imported into other software that can be used to build a reconstruction or to test ideas about how the animal functioned.

There are two major drawbacks to the use of CT scanning. The first is cost; a single scan can cost hundreds of dollars to perform and analyze and requires a highly trained technician. The second is that a strong contrast is needed between the fossil and the matrix in their transparency to X-rays. In my one attempt to have a fossil CT-scanned, I sent a crinoid specimen to a major facility in Texas. Unfortunately, the contrast between the fossil and its encasing sediment was too small for the scan to be successful. I did, however, receive a beautiful 3-D scan of the surface.

The next step up is a major one—using the X-ray beam from a synchrotron. The discovery of "synchrotron radiation" was serendipitous; when high-energy physicists used a curved particle accelerator on charges particles, they discovered that X-rays are emitted. It was quickly realized that rather than being a nuisance this provided a means to produce intense (bright) X-rays to study a wide range of materials. In very large accelerators, such as the Advanced Photon Source (APS) at Argonne National Lab in Illinois or the Stanford Synchrotron Radiation Lightsource (SSRL) in California, the X-ray beams produced are many orders of magnitude brighter than anything that can be produced by a conventional X-ray tube.[9] The result is a greatly enhanced ability to penetrate the matrix and produce CT-scans of the embedded fossils in incredible detail For example, many specimens of insects in amber are almost impossible to study under the light microscope because the amber is cloudy or full of bubbles. Using a synchrotron light source in Europe, Carmen Soriano and her colleagues were able to produce three-dimensional images of insects in amber in which the finest structures are visible.[10]

Where the synchrotron really excels is in telling us the chemistry of the fossil and the surrounding rock. When the X-rays strike the specimen, atoms are ionized and radiation in the form of a lower-energy X-ray is released; this is fluorescent radiation. The energy of the fluorescent X-ray is specific to the element present. As a result, a detector can determine what elements are present in the sample and in what quantities. By scanning across the entire specimen, a highly detailed map of the distribution of the elements can be made, an XRF X-ray fluorescence (XRF) image. This allows us to study the chemistry of both the fossil and the matrix.

In addition to the distribution of elements, synchrotrons can be used to determine the presence of compounds. The Pennsylvanian age (310 million years old) cave from Illinois that my colleagues and I studied yielded fragmentary remains of scorpions. We sent these specimens to colleagues working with the Advanced Light Source (ALS) at the Lawrence Berkeley Laboratory. They also examined 415-million-year-old cuticles of eurypterids. Using a method called XANES (X-ray absorption near edge structure spectromicroscopy; no wonder they use the acronym!), they were able to determine that the specimens preserved remnants of the chitin and protein complex of the original cuticles.[11] Previously, it was not thought that these molecules could be preserved so far back in time. The oldest chitin was then only known from Oligocene beetles (about twenty-five million years old).

The major downside of using a synchrotron for rock and fossil composition studies is that there are only a few such facilities in the world. In addition, highly trained technicians and scientists familiar with both spectroscopic methods and the nature of paleontological specimens are needed. Unfortunately, one such scientist, Carmen Soriano at Argonne National Labs, was let go, and the paleontology program was shut down even though it was producing amazing results. The real reasons are not clear.

For those without access to a synchrotron, many other methods can be used to determine material composition. Energy dispersive spectroscopy (EDS) using a scanning electron microscope has long been used to map the distribution of elements; I have used it to look at the elements in the shells of crabs. Raman spectroscopy sounds like it has something to do with noodles, but it is a method for looking at atomic bonds, including those in organic molecules. It involves looking at a very tiny fraction of the light that is scattered when a laser is shone on the sample; this scattered light has different wavelengths than the light source. The change is caused by the light's interaction with the vibrations of molecules in the sample.[12]

Yale graduate student Jasmina Wiemann is using Raman spectroscopy to look for preserved organic molecules in fossil vertebrate tissues; she considers herself a molecular paleobiologist. As part of her graduate work, she examined well-preserved Cretaceous dinosaur eggs and was able to show that they were colored and covered in spots and speckles.[13] This was the first demonstration of egg color in dinosaurs and the first time this been documented in a group other than birds.

A related method to Raman spectroscopy, in that it depends on the vibrations of molecular bonds, is infrared (IR) spectroscopy, which utilizes the absorbance of infrared light of different wavelengths. What is key for either approach is that the resulting spectra can be used as a fingerprint of the molecules that are present. For example, the same ancient cave that yielded the scorpion cuticle also contained extremely well-preserved spores of the plants that were alive then. A study of the wall of the spores with IR spectroscopy revealed that they were made of unaltered sporopollenin, the same resistant material that comprises the outside of their relatives living today.[14]

Some projects require multiple methods. One of the most remarkable episodes of Earth history are the pole-to-equator glacial events known as Snowball Earth. Occurring in at least two major episodes, 640 million and 710 million years ago, lasting millions of years, they define the Cryogenian period. The glacial deposits are topped with bizarre thick limestones known as cap carbonates. A major question is how did life survive these extreme episodes. Sara Pruss of Smith College, working with Tanja Bosak of MIT and their coworkers, addressed this problem using a range of advanced tools, including Raman spectroscopy, X-ray diffraction (XRD), and EDS, to look at organic materials and tiny fossils contained in the cap carbonates from Namibia and Mongolia.[15] They found that eukaryotic life survived and thrived. Sara says, "When and I first started this project, we were surprised at what a wondrous and tiny world existed in the residues of the cap carbonates. And then to find out the record was global confirmed that we were really onto something!"

This is only a fraction of the wide range of tools (and corresponding acronyms) used to study the structure and chemistry of fossils, including the discovery of ancient biomolecules.[16] There is also time-of-flight secondary ion mass spectrometry (ToF-SIMS), pyrolysis gas chromatography mass spectroscopy (Py-GC-MS), and my personal favorite acronym, electron energy loss spectroscopy using a scanning transmission electron microscope (STEM-EELS). I am sure I am missing some. In a later chapter, I discuss methods for studying ancient proteins and DNA. More important than these cool new toys themselves are the scientific questions they allow us to answer. An example of this is the identity of the state fossil of Illinois.

The state fossil of Illinois is the Tully monster, *Tullimonstrum gregarium*, from the Pennsylvanian age Mazon Creek. Not a Sesame Street character, it was named in 1966 by the late (and whimsical) Gene Richardson of the Field Museum after its discoverer, amateur fossil collector Francis Tully. One of the larger animals collected at Mazon Creek, it is remarkable for its lack of obvious resemblance to any living organism. Its original describers suggested it was a new phylum; it has since been compared to a wide range of animal groups. I have a couple of specimens in my office; occasionally, I would pick them up and think: "I should figure out what these are!" In 2016 two papers appeared together in the journal *Nature* that independently came up with a radically new idea on what *Tullimonstrum* was. One group, working mostly out of Yale and the Field Museum, used methods including digital photography, EDS using a scanning electron microscope, and synchrotron XRF spectroscopy at the Argonne Advanced Photon Source. They analyzed the entire body. The second group, primarily British scientists, focused on what was believed to be the eyes. They also used EDS, but in addition they employed ToF-SIMS, which gives the distribution of elements. The two groups of scientists came to the same and not uncontroversial decision: the Tully monster was a vertebrate and thus very distantly related to us.[17] This new interpretation could not have been made without the availability of new technologies and the willingness of paleontologists to use them.

The chemistry of fossils can also unlock the history of Earth's climate. Most marine fossils have shells made of calcium carbonate ($CaCO_3$), in the form of the minerals calcite and aragonite. Since the 1950s, it has been recognized that the oxygen in these shells can be used as sensitive indicators of ancient climate, especially of ocean temperatures. This takes advantage of the fact that there are two principal stable isotopes of oxygen, with atomic weights of sixteen (the common form, 16O) and eighteen (the less common form, 18O). As organisms make their shells, the slight difference in mass affects how the calcium carbonate is produced, changing the proportions of the two isotopes, a process called fractionation. This shift is sensitive to temperature, and shells produced in warm water have relatively more 16O and are thus lighter than shells produced in cold water.[18] The original standard with which these were compared was a fossil dubbed PDB (for Pee Dee belemnite); belemnites are extinct squid relatives, and the fossil was from the Cretaceous age Pee Dee formation of the Carolinas. The realization that analyzing the relative proportions of the two isotopes in fossil shells, most frequently those of the single-celled foraminifera, allowed the reconstruction of ancient temperatures in the oceans gave birth to the field of paleoceanography and the modern science of paleoclimatology. This research led to the recognition that there were not four ice ages during the Pleistocene but at least twenty and that they were strongly periodic. This result, in turn, led to the understanding that the ice ages were probably controlled by changes in the Earth's orbit. In a longer term, we now understand the slow cooling of climate that preceded the ice ages. As the micropaleontologist Ellen Thomas (Wesleyan and Yale University) indicates, "the curves of climate change over the Cenozoic are all derived from stable isotope analysis of microfossils." No discussion of anthropogenic climate change can take place without understanding this long-term natural history.

Another stable isotope pair that has become increasingly important are those of carbon, 12C and 13C. Again, using PDB as a standard, the carbon in plants is very light, a result of fractionation that occurs during photosynthesis. The ratio of carbon preserved in fossil material is thus a record of the global carbon cycle. One of the prime lines of evidence we have for the origin of life is the occurrence of very light carbon in rocks that are 3.5 billion years old. The ratios can be used indirectly to tell us how much oxygen is in the atmosphere throughout Earth's history.

Carbon isotopes also record a major shift in the world's vegetation. Most plants are called C3 plants because the first organic carbon compound made in photosynthesis contains three carbon atoms. This is contrasted with C4 plants, which use four carbon atoms. C4 plants include most tropical and salt marsh grasses, including corn and sugar cane. C4 and C3 plants differ strongly in their fractionation during photosynthesis. As it turns out, whether an herbivore eats C3 or C4 plants or a mix of the two is preserved in their tooth enamel. By looking at the fossil enamel, we can reconstruct the basics of their diet.[19] This has been a major line of evidence suggesting that C4 grasses spread during the Cenozoic to form modern grasslands.

Determining isotope ratios accurately required the development of sensitive mass spectrometers, equipment that measures the masses of atoms within a sample. In recent years, the precision possible using mass spectrometers has become mind boggling; they are able to measure the concentrations of isotopes at levels of parts per billion. This has opened a world of new isotopes to paleontology, including those in which the concentrations are quite low. For example, Michael Henehan, currently at the GFZ German Research Centre for Geosciences in Potsdam, has used isotopes of boron in foraminifera as a proxy[20] for ancient oceanic pH values, which in turn should reflect the concentration of CO_2 in the atmosphere (figure 6.3).[21] The boron concentrations in his sample solutions are on the order of fifteen parts per billion. His samples need to be prepared in a clean room fitted with special boron-free air filters to avoid contamination; workers and visitors must wear covers over their clothes and shoes.

Robin Canavan is a postdoc at the University of Massachusetts Amherst who examines periods of time in Earth history when the planet was usually

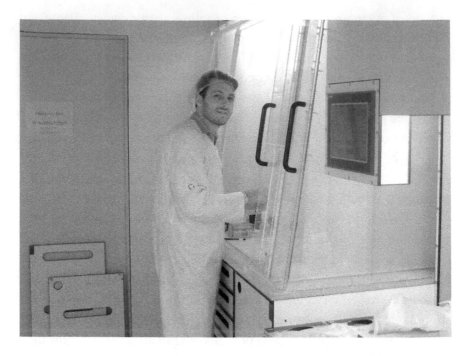

FIGURE 6.3 Michael Henehan at Yale University, studying boron isotopes in foraminifera.

Source: Photo by the author.

warm, the "greenhouse climates" (as opposed to glaciated "icehouse" like today). For this purpose, she is using a method called clumped isotopes to study the shells of mollusks. Clumped isotopes appear to be the hottest thing, so to speak, in paleoclimate research. The details are complex, but the method takes advantage of the fact that sometimes the rare isotopes of carbon and oxygen (13C and 18O) bond (clump) with each other in the carbonate minerals calcite and aragonite that make up many shells. The amount of these isotopes that bond is temperature dependent. This is a difficult method to use, requiring mass spectrometers of exquisite sensitivity.

The integration of geochemistry and biochemistry with paleontology shows, to a considerable extent, how porous and somewhat artificial disciplinary boundaries are. They are all part of unified effort to study the evolution of the Earth and the life on it. Pedro Marenco uses geochemical methods to study many periods of Earth history, such as the environment following the catastrophic end-Permian extinction. There are "many different ways to study the history of life," according to Pedro, "mine is to use geochemistry." Michael Henehan points out that getting reliable results from paleoclimate studies of isotopes in foraminifera depends on correctly identifying the species being used because each will respond differently to water temperature and chemistry. He is concerned that geochemists don't appreciate the complexities of biology: "micropaleontology and geochemistry cannot be separated!"

Finally, the advent of digital technology and computers has revolutionized another fundamental method in biology and paleontology: the measurement of the shape and size of living and fossil organisms. These measurements are an integral part of describing any new species, for comparing among a group of species, for examining how individuals grow, and for describing evolutionary changes within and among species. Documenting changes in size and shape are critical for studying how organisms respond to climate change.

The first step in this description is to make a measurement. One tool to do so that has a long heritage is a good pair of calipers; I used an analog dial caliper for my thesis work. I had to measure, write down the value, take the next measurement, etc. Now digital calipers can be directly connected to a computer (figure 6.4). For objects too small to measure with calipers, we had to use a special microscope eyepiece that was calibrated for each magnification. Again, measure, write down, repeat. New microscopes, like the one in my lab, automate the process and save each measurement into a file. Many of the imaging methods described here allow easy generation of measurements.

Once the measurements are made, we can analyze them. A huge number of sophisticated statistical techniques can be used, depending on the scientific questions being asked. The development of user-friendly statistical packages has made this easy. Many produce graphs that can be directly imported into papers. A favorite package for many is the Past program developed by Øyvind Hammer of the Natural History Museum, University of Oslo, which was specifically written for paleontological applications. Best of all, it is free![22]

The variety of technologically advanced methods used by today's paleontologist far exceeds what I have discussed here, but I think I have used enough acronyms. The image of the paleontologist's tool kit consisting of a hammer,

FIGURE 6.4 Neal Landman (American Museum of Natural History) is measuring Cretaceous age ammonoids using digital calipers.

Source: Photo by the author.

a whisk broom, and a hand lens is seriously flawed. Instead, we are incredibly eager to use any technological innovation that will make our job easier and produce more information. The low-tech stuff is still used and just as essential as ever, but it is now supported by a plethora of cutting-edge analytical methods. We have cool toys and cool ideas!

7

BIG DATA AND THE BIG PICTURE

As an undergraduate, I was fortunate to have a work-study job with Niles Eldredge at the American Museum of Natural History. He was describing a beautifully preserved horseshoe crab from the Devonian of Bolivia. Niles assigned me the task of compiling a bibliography of fossil horseshoe crabs. Going to the museum library, I found the volumes of the *Zoological Record* that covered the chelicerates, the arthropod group containing the horseshoe crabs. The *Zoological Record* was a huge annual publication that listed all new species names that had been published that year. I also used the *Bibliography and Index of Geology*, which recorded every paper published on a geological subject, including fossils, for a given year. I photocopied the pages listing the papers on horseshoe crabs, cut each reference out, and glued it on an index card. The result, after weeks of effort, was a small file box with nearly all the publications on fossil horseshoe crabs.

I used the same approach to compile the bibliography for my PhD with the late J. John (Jack) Sepkoski. One goal of my thesis was to document the history of the eurypterids (sea scorpions), an extinct group of scorpion and horseshoe crab relatives that included the largest arthropods of all time. Using the papers in the card file, I recorded on index cards what species were found at each locality and, on a separate set of cards, which localities had which species. This information was then combined, usually on graph paper, to produce a synthetic picture of the diversity of the group over time. The entire process was tedious and time-consuming.

I could carry out the same procedure from scratch today in a small fraction of the time. Bibliographic information on nearly every scientific paper is now available for download from online databases, including the *Zoological Record* and the *Bibliography and Index of Geology*. Entering the search terms "fossil" and "horseshoe crab" produces scores of references within seconds that I can save in my desktop reference manager. If my university library has paid the journals for access, I can then download the papers themselves almost as quickly. Publications prior to the digital age are also often available as scanned copies, although the quality of the scans frequently leave something to be desired. I can then enter the data from the papers onto spreadsheets or other database programs, plot my results using a statistics package, and produce publication quality figures with sophisticated graphics software. This enables me to spend my time thinking about the science and writing rather than doing grunt work. In this case, I have absolutely no nostalgia for the analog age.

In addition to chapters on the distribution of eurypterids in time and space, my thesis included chapters on the shape variation within and among eurypterid species, using statistical methods on a computer; on the relationships among the species, using a method called cladistics; and an experimental study of their swimming abilities. What it did not have was descriptions of new species or localities, the meat and potatoes of traditional paleontological theses. My doctorate thus reflected the transformation of paleontological research that occurred from the late 1960s to the 1980s, a period of intense ferment that has been labeled the "paleobiological revolution."[1] Borrowing two terms from psychology, Steve Gould in 1980 urged that paleontology transform from an "idiographic" discipline that only examined "unique, unrepeated events," such as the occurrence of a species at a locality, to a "nomothetic" field focused on "the lawlike properties reflected in repeated events";[2] that is, the overall patterns and their controls. This transformation was the explicit goal of the paleobiological revolution.

The recognition of large-scale patterns in the history of life—the "Big Picture"—long predates the revolution. By the middle of the nineteenth century, it was clear that definite patterns in the fossil record were recognized in the emergence of both regional and global geologic time scales. Most conspicuous were the divisions of time into the Paleozoic (old life), Mesozoic, (middle life), and Cenozoic (new life) eras, terms dating back to 1841, that recognize fundamentally distinct types of organisms living in the oceans and on land. Development of the time scale also reflects the first attempts to quantify life's history. Charles Lyell subdivided the classic sequence of Cenozoic rocks near Paris into the Pliocene, Miocene, and Eocene. These epochs, as we now call them, were based on the percentage of mollusks (clams and snails) alive today that they contained: 90 percent of living mollusks were found in Pliocene rocks, 18 percent in the Miocene, and 9.5 percent in the Eocene.

Until the late twentieth century such synthetic and numerical approaches to paleontology had been only a small part of the discipline. A dramatic change in this was presaged by the 1940s and 1950s writings of the great vertebrate paleontologist George Gaylord Simpson. His books, *Tempo and Mode in Evolution* (1944) and its greatly revised version *The Major Features of Evolution* (1953), remain required reading for current generations of paleontologists. These books asked important questions: How fast does evolution occur? Are there trends in evolution? How long do species survive? Do rates and trends and species survival differ among different groups of organisms? Are their differences between the evolution of species and that of larger groups such as families? He also asked profound questions about the processes, such as adaptation, that drive the evolution of species and higher groups. Simpson's books are full of tables and graphs and contain only a bare minimum of fossil sketches. In the language of Thomas Kuhn, Simpson established the paradigm for the current discipline of paleobiology.

Simpson's work, and that of Norman Newell at the American Museum of Natural History, laid the groundwork for the paleobiological revolution. I first became aware of the ongoing change in the discipline when I took a graduate course with David Raup at the University of Rochester. It was a paleontology course without fossils![3] The syllabus included topics such as "Markov processes," "Monte Carlo techniques," "idiographic and nomothetic approaches," and "taxonomic ratios and the hollow curve." Rather than dealing with descriptions of the morphology or stratigraphy of fossil organisms, it focused on techniques for describing and modeling the broad-scale patterns in the fossil record. One of the key concepts Raup introduced to us (and the rest of the paleontological world) was the idea of the "null hypothesis," a concept derived from statistics that asked us to rule out random causes for patterns before invoking deterministic explanations. This idea is exemplified by the title of one of Dave's highly readable books, *Extinction—Bad Genes or Bad Luck?*

Once rare, these numerical approaches are now ubiquitous in the field. Any recent random issue of the journal *Paleobiology* may include only one or two articles with a picture of a fossil; most of the remaining papers are filled with graphs and numerical formulas of different degrees of complexity. The same is true of many sessions at our annual conventions—hours may go by without a single illustration of a fossil. A cartoon on my door shows a man and a woman seated in front of computer screens; the man turns to the woman and says, "Remember when geologists worked outdoors?" This rings true for numerous paleontologists too; many of us spend far more time in front of a computer than we do in the field or in the laboratory.

Part of this transformation is intellectual; the paleobiological revolution and its aftermath (no pun intended) profoundly changed the nature of the questions that were asked about the fossil record. David Jablonski (University of Chicago) put it this way: "If I had to say just one thing on how things have changed, it's the influx of quantitative approaches (not just methods but a whole way of thinking) . . . when I started in the early 1970s paleontology was cast much more in the handmaiden role and only a few people were addressing big questions." Derek Briggs of Yale concurs, saying that when he started in the 1970s, British paleontology was focused on taxonomy and systematics (description and classification); the discipline is now far more question oriented. As Peter Wagner (University of Nebraska–Lincoln; box 7.1) puts it, analytical paleontology "let us become full-fledged hypothesis testers for large questions . . . it gave us the 'null hypothesis.' "

BOX 7.1

Peter Wagner remembers getting interested in fossils when he received a book about dinosaurs from his grandparents, although he was also interested in astronomy. As an undergraduate at the University of Michigan, he did not take paleontology because it was taught in geology and he was a major in biology and physical anthropology. When he took anatomy, however, he realized that all "the cool stuff was fossils." He decided to get a second bachelor's degree, in geology this time, at Michigan State. Peter stayed on at Michigan State to get his master's, working with the paleobiologist Doug Erwin (now at the Smithsonian). He then went to the University of Chicago for his PhD. By academic genealogy, Peter Wagner can be considered my "younger brother." Although not literally true, we both had the same PhD advisor in the late Jack Sepkoski.

The other driver of the transformation was technological; as in so many areas of science, the advent of fast and cheap computers and the internet have fundamentally changed how paleontological research is conducted. The internet has led to the development of large online databases of paleontological, geological, and biological data. And fast computers provide the means for highly sophisticated analyses and models for that data. George Mason University paleontologist Mark Uhen says, "everything we do with computers we could do without them, but having them is much better." Paleontology has entered the era of what has been called "big data."

Much of this research has focused on major questions about the history of life, sometimes called "macroevolution." First, how many kinds of organisms were alive in each interval of geological time; that is, what is the history of biodiversity?[4] Second, how do the magnitudes and rates of originations of new forms and the extinctions of existing ones change over time? This includes the study of episodes of unusually high rates of extinction (mass extinction) or origination (evolutionary radiations). In addition, how do these numbers, rates, and magnitudes differ among various kinds of organisms? These questions are not independent. The number alive at any time is the sum of how many survived from past times, how many new forms appear, and how many go extinct. Underlying all of these questions is estimating how faithfully the observed fossil record preserves the original history of the biosphere; that is, what is the fidelity of the fossil record? All of these questions of number, magnitude, rate, and fidelity must be addressed before we can reliably answer the more interesting question of the mechanisms that produce the patterns.

The origins of big data in paleontology are not tied, as it is in other areas of the earth sciences, to the development of high-tech equipment, such as satellites that can generate terabytes of data. Instead, it derives from the long history of the field and the patient effort over hundreds of years to describe fossils and their localities. The descriptive literature of paleontology is quite literally the ground truth for all theoretical explanations of the history of life. It plays the same role that observational astronomy plays to cosmology or experimental physics plays to theoretical physics. The importance of literature is also why paleontologists value libraries and old books and journals; the information in them is still relevant, even if was published a century ago.

To be used, the wide and scattered literature of paleontology must be brought together. Beginning in the early 1950s, the *Treatise on Invertebrate Paleontology* published summary volumes on all of the nonvertebrate fossil groups. Still not completed, it currently numbers more than fifty matching volumes, each with hundreds of pages (it has recently begun to appear online). Most important, it summarizes the fossil record of genera and families in those groups, including when they first and last appear: that is, their ranges. Similar data for vertebrates can be found in the exhaustive textbook written by A. S. Romer and similar sources. Working with these books, and an earlier compendium edited by W. B. Harland, Jack Sepkoski Jr. produced "A Compendium of Fossil Marine Families" in 1982 that summarized the ranges over time of some 3,300 families of ancient organisms. This compendium was used in numerous studies by Sepkoski and his coworkers, in particular David Raup, to summarize the patterns shown by the data within it, including a still controversial attempt to show that mass extinctions occurred periodically[5] and one paper by Sepkoski that included a now iconic diagram on the overall history of life's diversity.[6] A version of that diagram still appears in some talks at every major conference.

Jack was not an expert on most of the groups contained in the compendium, and he was heavily criticized for errors in it by those who were specialists. As a result, he spent the next "ten years in the library" updating the data set, which he published in 1992.[7] Although about half the original data was changed, the overall patterns of life history were not substantially altered, suggesting that the signal they represented was very strong compared to any noise in the data. These include such events as the Cambrian Radiation (aka the Cambrian Explosion) and the major mass extinctions. (I will return to these.)

Families are made up of many genera and even more species, and they are a fairly crude way to measure diversity over time. Next Sepkoski turned his attention to genera, a much more intensive project. By the time of his death in 1999 at the age of fifty, he had compiled a list of some 37,000 fossil genera and their first and last occurrences, which was published posthumously in 2002. The resulting volume is 560 pages long! *A Compendium of Fossil Marine Animal Genera* formed one nucleus of the next critical step—creation of the Paleobiology Database.

Sepkoski's database, although invaluable in many studies and representing many years of effort, still had admitted shortcomings that reduced its usefulness. To begin with, the database omitted terrestrial organisms and marine organisms other than animals. There was no information on species. Nor was there anything about their ecology, geographic distribution, abundance, preservation, or any other geologically or biologically relevant information. Critically, it lacked any geologic data on a genus other than its first and last occurrence, constituting what is called "range-through data," without any indication of how often it might be found between these times, if at all. Further complications occur because different intervals of Earth history are greatly dissimilar in their durations and their quantity of sedimentary rocks; the longer time and the more rocks there are in an interval, the more fossil genera will be counted from it. These, and other factors, strongly suggest that simply counting the number of genera (or species or families) in an interval gives an inaccurate picture of diversity over time. To correct this requires sufficient amounts of the right kind of data. In particular, this means going back to the original source material of the discipline, the work of generations of paleontologists reporting what they had discovered in the field.

To address these issues, in 1998 Charles Marshall and John Alroy, with assistance from Arnie Miller, began development of what became known as the Paleobiology Database (PBDB). The PBDB had its roots in several preexisting databases and compilations.[8] Since then, it has grown to become an essential part of the infrastructure of paleontology. It is an enterprise fueled by the entire paleontological community, not just a single individual. Working scientists were actively engaged with setting up the structure of the database, so it reflects our knowledge and the scientific questions we want to answer. The PBDB has strong community support and participation. As of July 2019, data had been entered by 410 scientists from over 130 institutions in twenty-four countries. It has also fissioned; the original site and its analytical tools continues to be managed by John Alroy at Macquarie University in Australia under the name *Fossilworks*; a newer site with a different set of tools for data download and analysis using the title *The Paleobiology Database* is housed at the University of Wisconsin, under the leadership of Shanan Peters. Both sites contain the same data sets.

The data in the Paleobiology Database is fundamentally different from that developed by Sepkoski. The focus is on described fossil collections; that is, a list of fossils from a specific location (an "occurrence") representing a single geologic unit, usually one that has been described in the paleontological literature. For each collection, in addition to the list of fossils present, there are precise geographical coordinates, a detailed description of the associated rock units with an interpretation of their environment, and the best available estimate of the age of the rocks. In addition, information on how the fossils are preserved can be entered and, if available, numbers on how common each kind of fossil is. The goal is to capture all of the information contained in the original author's description of the collection as well as any information added by later authors. Parallel to the data set of localities is one that contains the current knowledge of the classification of the fossils at the locality.

Entering all of this information can be time-consuming; information on even a single paper can take several hours to enter, depending on its level of detail. But the amount of data in the PBDB is astonishing! As of July 2019, information from more than 69,000 papers had been entered describing 202,000 separate fossil collections and 355,000 different named taxa. Combining all species lists yields about 1,420,000 occurrences in the database. It may not be terabytes, but for us it is indeed big data! Especially exciting is that efforts are ongoing to integrate the PBDB with other databases that document distinct aspects of Earth history, enabling us to truly look at the history of the entire Earth system.

Equally important to the size of the PBDB is its accessibility. It is a public database, meaning that it is freely accessible to all paleontologists, all scientists, and anyone else with an interest in the fossil record. You can go to the site and find the paleontological information you need. No subscription, no firewall, no fees.

What makes the PBDB so useful is that these data are very easy to download, and you can be very specific in what kind of information you want. The flexibility with which data can be downloaded is terrific, ranging from simple menu-based queries to highly sophisticated searches via what is called an "Application Programming Interface." You can download information on a single locality or taxon or on defined groups of localities or taxa, and you can refine your request by a host of conditions, such as age, geographical region, or environment of preservation. For example, if you enter *Triceratops* as a search term, you will get a page that tells you the current classification, provides a list of references on the dinosaur genus, gives you a simple summary of its ecology, and provides an overview of the distribution in time and space of the 136 collections containing *Triceratops*. If you click on the links in the distribution area, you will see one of the localities where *Triceratops* has been found. And there you will find all the original information on the age, location, and geology of the location, where it was described, as well as a list of all the species that have been collected there. To put it simply, the PBDB is the starting place for any exploration of the known fossil record.

Learning to work with the PBDB has become part of the training of new generations of paleontologists. Beginning in 2005 under the guidance of John Alroy at the Fossilworks Intensive Workshop in Analytical Palaeobiology, and continuing with a short hiatus until today, the month-long Analytical Paleontology Workshop has taught scores of young paleontologists from across the world how to work with the database and the wide range of analytical methods used in modern paleobiology.[9] One of the current instructors,[10] Phil Novack-Gottshall of Benedictine University, describes the goal this way: "Because most paleobiologists don't have a formal background in computer programming, the month-long course helps graduate students gain the programming experience and conceptual know-how to carry out their own analyses." The programming is in the R language, which has become the standard in many areas of science. Phil adds that "students are really enthusiastic about the course, very highly engaged in learning the dense material in a short period of time (they only have one free day per week!), and ready to incorporate the techniques into their research."

For paleontologists, the PBDB has become an essential resource. As of July 2019, the project itself was nearing 350 official publications across a wide range of topics, as well as about 25,000 papers across geology and biology that cite use of the database. In my own research, I have used data in the PBDB in multiple papers, covering a wide range of topics. For example, I was curious about which kinds of fossils are the most common; that is, which ones are most likely to be found in the field by a geologist or paleontologist. Working with my colleague Peter Wagner, we used the PBDB to determine that a very small fraction of the thousands of genera is responsible for most of the occurrences in the database.[11] These, in almost all cases, were the same genera that you find in every introductory paleontology or historical geology course laboratory. The most common genera were also the most familiar.

Ultimately, the goal of many researchers using the PBDB is to address the big picture questions about the history of life.[12] Having a huge amount of information enables us to use highly sophisticated numerical methods to correct inherit errors and biases in the record so that the evolutionary signal can be clearly seen. As a result, there is a growing if not complete consensus on the overall history of biodiversity, of life in the oceans since the beginning of the Cambrian. And there is also enough data to start to answer the big question of what produced these patterns; not just the what but the why of the history of life.

But I issue a caveat: the PBDB is only as good as the data that is put into it. Several paleontologists have pointed out significant errors in how their own group is represented (although they are welcome to participate and make the corrections). In a recent paper, I was using the latitudes and longitudes of fossil localities as the input to my analyses. A reviewer pointed out that many of the points I had planned to use were unreliable because of how they were entered. So warned, I was able to omit them and redo the study.

Even more important is that the growth of paleobiology has been paired with a loss of expertise in descriptive paleontology, the fundamental data on which the entire enterprise depends. There are entire areas of the tree of life that lack an expert to validate the information in the databases. As a result, some paleobiological papers that use highly sophisticated and impressive sounding methods are useless because the authors do not understand the nature of the data they are analyzing. The science needs both the talented user of R-code and the describer of the hinge area of productid brachiopods.

Paleontological data alone is not sufficient to address the big questions. Their resolution depends on the use of information from other areas of science and the big data sets that they are also developing. Answering the big questions also depends on teamwork that cuts across and unites disciplines, with paleontologists acting as a bridge. Two examples of this multipronged approach are research into the origin of animals and the timing, scales, causes, and consequences of the major episodes of extinction.

Since the origin of life, some 3.5 billion years ago (the exact date is furiously debated), nearly all living things on Earth have been single-celled organisms. First to appear were the bacteria and archaea, collectively known as prokaryotes. Small, lacking a nucleus containing their DNA, they also did not require oxygen for their metabolism. Some of them were photosynthetic and produced oxygen as a by-product. About 1.6 billion years ago, a more complex cell appeared. These eukaryotic cells were larger, had their DNA bundled in a nucleus, had organelles such as mitochondria, and were dependent on the oxygen that had begun to accumulate in the oceans and in the atmosphere. But they were still single cells. It was almost another billion years before multicellular life appeared, about 565 million years ago, in a period called the Ediacaran (box 7.2). The Ediacaran organisms are enigmatic; their relationships with each other and with later life have been debated since they were first discovered. Following the Ediacaran is the Cambrian Explosion, a period of time starting 541 million years ago and lasting for about 20 million years, during which essentially all of the major animal groups, the phyla, first appear in the fossil record.[13] These phyla include the vertebrates; to which we belong. It is also when the familiar fossil record began, as many of these organisms also made shells and other hard parts. By the way, I am among those who dislike the term "explosion," with its false implication of suddenness. I prefer the term "Cambrian Radiation," which is admittedly less sexy but more accurately conveys that this was not an overnight event.

BOX 7.2

Lidya Tarhan never grew out of her childhood interest in paleontology, although she had many others, including archeology and the humanities. As a geology and English double major at Amherst College, she received a fellowship to work out in the field with then Amherst faculty member James Hagadorn. She was hooked. At the University of California Riverside, Mary Droser gave her the opportunity to work on the amazing Ediacaran rocks in Australia, which record some of Earth's earliest multicellular organisms and have been the focus of much of her subsequent research. Lidya is now an assistant professor at Yale University.

Understanding what happened during the Ediacaran and Cambrian periods has been the focus of intense research for decades, with numerous competing explanations for why multicellular organisms appeared when they did and why they diversified so comparatively rapidly.[14] These investigations have necessitated the joint efforts of paleontologists, geologists, geochemists, and biologists. Paleontologists have used data sets such as the PBDB to determine the times and rates of origin of major groups of animals and describe new organisms that illuminate the dynamics of the radiation. Geologists have used high-resolution methods to determine the exact timing of the events; for example, the resolution on the 541-million-year-old boundary between the Ediacaran and the Cambrian is about 0.1 million years. Geochemists have employed a multitude of sophisticated methods to constrain what we know about the chemistry of the oceans and the atmosphere over this interval. Critically, estimates for the amount of oxygen present at the beginning of the Cambrian are on the order of 15 to 20 percent of that in the modern atmosphere.[15] Before that, oxygen levels were probably much lower and variable from place to place. Biological studies indicate that these levels were sufficient to maintain multicellular life (figure 7.1). Biologists have also used the DNA of modern animals to determine how they are related to each other and when the various groups may have appeared; the latter is dependent on the fossil record for calibration. Although there is a great deal of controversy and uncertainty, this "molecular clock" research suggests that the ancestors of living animals first diverged in the Ediacaran, well before their appearance as fossils. Finally, studies of the genetics of development and of ecology indicate that evolution of new genetic interactions within organisms and ecological interactions among organisms were made possible by the rise of oxygen. In sum, paleontologists provide the hard evidence for "when" animals first appeared, and through collaborative efforts with other sciences we can come to understand the "why" of it.

A similar story of synergistic interactions of other fields is in the study of the large mass extinctions that have dramatically punctuated the history of life on Earth. Extinctions are always occurring, the so-called background extinctions, but at certain times in Earth's history large numbers of extinctions occurred in a relatively short time. These episodes are the mass extinctions, which have been recognized in one form or another since the earliest days of the field of paleontology. Describing, comparing, and determining the causes

FIGURE 7.1 The trace fossil *Treptichnus pedum*, made by an unknown animal. Its first appearance marks the beginning of the Cambrian. Grand Bank, Newfoundland.
Source: Photo by the author.

for these dramatic events have been a preoccupation with paleontologists ever since and remain a source of fundamental disagreements to this day. Probably the best-known summary of extinctions since the beginning of the Cambrian was published by Jack Sepkoski and Dave Raup in 1982, based on data in the Sepkoski database. Their analysis suggested that there were five major mass extinctions, at the ends of the Ordovician, Devonian, Permian, Triassic, and Cretaceous, the last of these being the one that claimed the dinosaurs (except the birds, of course).[16] These Big Five mass extinctions have now reached iconic status, with the current destruction of global biodiversity named the Sixth Extinction. Again, the advent of the Paleobiology Database and similar large data sets have greatly refined our knowledge of the scale, timing, and dynamics of these extinctions and even raised the issue of whether there were really five.[17] (I return to extinctions in the next chapter.)

One central message of this chapter is that without paleontology we would not know how the modern biosphere came to be. We would not know that the world of today is just a single snapshot of a long history. We would not know what the ancestors of modern animals looked like and how they lived. We would not know how changes in climate and geography caused by plate tectonics influenced the distribution of plants and animals. And we would not know how life survives and recovers from the very worst of natural disasters.[18]

Another message is that paleontology is an integral and inseparable part of the biological and geological sciences. Paleontologists are among the most multidisciplinary and interdisciplinary scientists. They work with geochemists, geochronologists, stratigraphers, biogeographers, systematists, geneticists, ecologists, and pretty much every "ist" that is out there. And they then can integrate among all of them.

A final message is that we need to banish the image that paleontologists (in the detested phrase) are stamp collectors. Instead, paleontology is increasingly a theoretical, hypothesis driven science that relies on the fossil record as critical data to frame our questions. We use big data, on big computers, to ask big questions. And then we go outside and find trilobites.

8

THE ENDS OF THE WORLDS
AS WE KNOW THEM

The end of the Cold War and the fall of the Soviet Union pushed the potential horror of nuclear war out of our thoughts, but it did not end our fascination with world-ending devastation. Postapocalyptic fiction remains a fixture of our cultural landscape, whether in books such as Margaret Atwood's *MaddAddam* trilogy or in movies like the Mad Max series. Climate change or bioweapon-caused pandemics have replaced nuclear war as the preferred agents of destruction, but the underlying cause is still the capability of humans to destroy themselves. What this fiction generally overlooks is that these catastrophes would be just as devastating for the living things that share the planet with us as it would be for humanity. Fictional or not, the ability of the natural world to survive and recover from an environmental disaster would be critical for our own survival. It is essential to understand what happens to the biosphere when global environmental catastrophes occur, and no one understands this better than paleontologists.

The recognition that extinctions happen is one of paleontology's great contributions to human thought, dating back to the eighteenth century when Georges Cuvier proved that mastodons no longer walked the Earth. Not long after that, as paleontologists documented the history of life on Earth, they recognized that extinctions came in waves of mass extinctions, with the largest episodes killing most of the species on Earth. The study of mass extinction—how many there were, how large they were, and what caused them—is a major focus of research not only in paleontology but in many other areas of earth science. It is also a topic of great relevance today.

In the early 1980s, the asteroid impact theory for the end-Cretaceous extinction and the documentation by David Raup and Jack Sepkoski of four other major marine mass extinctions in the last 540 million years brought renewed attention to the fossil record of biodiversity loss. Biologists working on modern biodiversity declines were paying attention. Writing in 1985, the renowned biologist Edward O. Wilson said, "The rate of extinction is now about 400 times that recorded through recent geological time and is accelerating rapidly. Under the best of conditions, the reduction of diversity seems destined to approach that of the great natural catastrophes at the end of the Paleozoic and Mesozoic Eras, in other words, the most extreme for 65 million years."[1] Within a short time, the current loss of species was dubbed the Sixth Extinction, a term that rapidly gained wide usage.[2] This concept reached public attention with the 2014 publication of Elizabeth Kolbert's Pulitzer Prize-winning book, *The Sixth Extinction: An Unnatural History*.

Can we better understand the putative Sixth Extinction through the mirror of the other five?[3] The fundamental issue is how comparable the modern biodiversity decline is to the catastrophic collapses of the geologic past. This leads to a whole series of basic questions about the past episodes: Who were the victims? How fast did the declines occur? What was the primary trigger for extinction, and how did it cascade through the biological system? How does life recover after extinctions, and how does the rebuilding biosphere differ from what went before? Another underlying issue is how the data to assess extinctions used by biologists compares to that available to paleontologists. The best understood mass extinctions are those at the terminations of two great eras of Earth history: the end-Permian extinction at the finale of the Paleozoic and the end-Cretaceous extinction at the conclusion of the Mesozoic.

Much has been argued about the extinction at the end of the Cretaceous, which goes by the shorthand K-Pg extinction, where K is the geological symbol for the Cretaceous Period and Pg that for the succeeding Paleogene. These controversies go back to a 1980 publication suggesting that an extraterrestrial impact caused it and the 1991 discovery of the impact site in the Yucatan.[4] Although the reality of the impact is generally accepted, fierce disagreements still exist over whether it was the sole or even perhaps the major cause of the disaster that befell the biological world. There are, for example, contemporaneous major volcanic eruptions in India.[5] I tend to strongly favor the primary role of impact and have since it was first proposed; recent studies strongly support this.[6]

What happened on that terrible day sixty-six million years ago? To answer this question requires using cutting edge field and analytical methods and drawing on the expertise of large numbers of paleontologists and other earth scientists.

In 2016, multiple cores were drilled into the 200 kilometer diameter Chicxulub crater left by the impact. In the wonderfully titled paper, "The First Day of the Cenozoic," a large team of scientists go step-by-step through the minutes and hours following the impact, documented by 130 meters of rocks filling the crater.[7] At the impact site, the initial stages were dominated by the formation of the 600 to 1,000 meter deep crater. The impact produced a huge tsunami, which propagated across the ancestral Gulf of Mexico, inundating shorelines throughout the region and then being deflected into the crater, filling it.

Being far from the impact would not have helped. At this time, what would become North Dakota was along the shores of an elongate inland sea, the last remnant of the Western Interior Seaway that once covered much of North America. The force of the impact and the resultant earthquake almost instantaneously generated a wave surge, a seiche, within the sea. As described in a highly publicized paper, fish and other organisms living along the shores were catastrophically buried.[8] Even more amazing, their gills contained tiny spheres of material ejected from the impact. They died that day!

If the tsunami and the seiche was not bad enough, the heat released by the impact set fires within 1,500 kilometers of the impact. As the debris from the impact rained back down, it would have also been hot enough to set numerous additional fires. Soot from the fires would reach high into the atmosphere, shutting down photosynthesis across the world, leading to starvation of herbivores and the carnivores dependent on them.

And it gets worse. The rocks where the impact hit were formed by evaporation when the Atlantic opened millions of years earlier; these rocks were rich in minerals containing sulfur. This sulfur was released into the atmosphere by the impact, producing sulfuric acid, which would have come down as acid rain, dropping the pH of the oceans. The rapid acidification of the oceans and a resulting huge decline in marine primary productivity were probably the main causes of the marine ecosystem collapse.

Tsunamis, seiches, fires, darkness, acid rain, and oceans. Overall, it was a terrible, horrible, no good, very bad day, even in Australia (apologies to Judith Viorst).[9]

The K-Pg extinction is best known because of the extinction of the dinosaurs ("nonavian dinosaurs," not my favorite term[10]) except for birds, although the birds did suffer dramatic losses. Extinctions struck plants and insects. Mammals may have lost more than 90 percent of their species. In the sea, the enormous marine reptiles, as well as the elegant, shelled cephalopods know as ammonites, disappeared forever. The tiny floating single-celled planktonic foraminifera, whose calcareous shells preserve key signals of ancient climate, virtually died out; out of seventy species, only three survived.[11]

Following the impact and its aftermath, life gradually began to recover. At the impact site itself, it is estimated that it took about thirty thousand years for a high-productivity ecosystem to be reestablished in the area.[12] At other sites in the Gulf of Mexico, it may have taken as long three hundred thousand years to recover. Far from the site, in the Pacific Ocean, ecosystems appear to have stabilized about 1.8 million years following the mass extinction, but phytoplankton diversity appears to have taken about eight to ten million years to recover to preextinction levels.[13]

How about on land? A remarkable section in Colorado records the first one million years of the Paleocene.[14] The sequence contains remarkably complete remains of plants, reptiles (crocodiles and turtles), and mammals. The first one hundred thousand years postextinction was a time of ecosystem instability, with full recovery taking about three hundred thousand years.

The K-Pg extinction remade the biological world. Life survived, but it would never be the same. To a large extent, the diversification of mammals and the eventual spread of humans was contingent (as Steve Gould would say) on that devastating event. Life on land and in the oceans today is very much the product of the recovery of the biosphere from the disaster. This is again where the integration of paleontological data with that from modern biology comes into play. A case in point is the evolution of birds and mammals. Early birds date back to the Jurassic, more than one hundred million years before the end of the Cretaceous. The first mammals date back even further, roughly contemporaneous with the first appearance of the dinosaurs in the Triassic. It was not until after the K-Pg extinction, however, that the diversification that produced nearly all our living bird and mammal groups occurred.[15] By ten million years later, there were bats, whales, and almost everything in between. The combination of molecular and paleontological data suggests that once the dinosaurs disappeared, the pace of appearance of new mammal and bird groups was extraordinarily rapid. The K-Pg extinction gave birth to the modern biosphere.

The end-Permian (P-Tr for Permian-Triassic), 252 million years ago, was the largest extinction in the history of life.[16] It has been called the "mother of mass extinctions," "when life nearly died," or "the great dying." Overall, estimates of the total loss of marine species range from 81 percent to 96 percent.[17] Many groups that were typical of the Paleozoic (rugose and tabulate corals, the already uncommon trilobites, and blastoid echinoderms) disappeared altogether, and many others were greatly reduced in diversity (brachiopods, crinoids, and ammonoids). Understanding what happened and why it happened has again required cooperative work from many kinds of earth scientists working together with paleontologists.

Instead of an impact, the trigger was massive volcanic eruptions. Geologists have found huge ancient volcanic deposits of the right age in Siberia, the Siberian Traps. The eruptions had multiple phases lasting for more than one million years. The onset of the volcanism probably produced major climate shifts. Concurrent with the marine extinction at 251.9 million years ago, the lava then intruded rocks rich in organic matter, releasing vast amounts of the greenhouse gases, carbon dioxide and methane, into the atmosphere. This led to a sudden large increase in atmospheric and oceanic temperatures, perhaps reaching lethal levels, and acidification of the oceans.[18] There is also evidence that the oxygen levels in the ocean dropped dramatically, and the ozone layer may have vanished. The rapid onset of extreme environmental change led to the collapse of marine ecosystems, including severe drops in primary productivity. So things went very bad, very fast.

When and how fast did it happen? A study in 2014 examined a geologic section in China.[19] Using multiple lines of evidence, they established with astounding accuracy and precision that the marine extinction took place

251.9 million years ago, over an interval of some sixty thousand to one hundred thousand years. On a geologic scale, this is an eye blink.

On land, the picture is less clear; relatively few places have rocks that preserve the critical interval. For land plants, there is surprisingly little evidence for a global mass extinction,[20] but the typical flora of high latitudes collapsed some 370,000 years before the mass extinction in the oceans, perhaps due to the initial climate disruptions from the volcanos.[21] It is also estimated that 70 percent of terrestrial vertebrates disappeared, but increasing evidence suggests that their extinction may have occurred about three hundred thousand years before the marine pulse.[22]

And how fast did life recover? Environmental disruptions, probably linked to continued volcanism, continued to suppress diversity and ecosystems. Marine temperatures remained hot and productivity remained low throughout the Early Triassic, and it was about five million years before marine systems began to recover.[23] Corals, produced by a different group of animals than the extinct Paleozoic forms, did not reappear until the Middle Triassic. Neither did the reefs they formed. Ammonoids rebounded. Ocean life became more familiar with the growing dominance of bivalves and snails. On land, the surviving reptiles gave rise to the first dinosaurs and mammals. Life survived, but the biological world was forever changed.

Although the external triggers for the K-Pg and P-Tr mass extinctions were different, several key lessons can be drawn. First, they produced rapid changes in the chemistry and temperatures of the air and water. Second, these extinctions tend to be "bottom-up," with the collapse of primary production cascading up the food chain, leading to the demise of numerous species in a relatively brief amount of time. Third, although life is resilient, it can take millions of years for ecosystems to return to some resemblance of preexisting conditions. Finally, to a large extent, our modern biosphere is the product of mass extinctions, who lived and who died. The study of ancient extinctions highlights an important fact: paleontologists are uniquely positioned to know what happens when severe environmental change stresses life. This gives us a distinctive perspective on the role of human-produced global change on the natural world.

How does the Sixth Extinction compare? First and foremost, the trigger is not external to the biological system but is very much a part of it. The numerous anthropogenic stresses on animals and plants, such as hunting, overfishing, the introduction of invasive species, habitat destruction, and climate change, are well documented. So is the ongoing loss of biodiversity, which consists not only of the extinction of species but also severe reductions in the population sizes of those that survive, a process called defaunation.[24] We have replaced the wide diversity of wild plants and animals with a limited number of domesticated species, to the extent that some 90 percent of all mammalian biomass are humans, their pets, and their livestock.[25]

The structure of the Sixth Extinction is also fundamentally different. Instead of being bottom-up, it is largely top-down. Although we have replaced native vegetation with agriculture, overall primary production on land has not changed. Mammalian biomass has increased as more and more plants are used to feed livestock. What we have lost in recent centuries are large native animals, including predators. Bison, caribou, elk, bears, and wolves were once common where I live; they are long gone. Humans are "super predators," hunting and fishing not for the weak and the old but for the large and the strong that will produce the next generation.[26] We are also "super dispersers," introducing our plants, animals, and diseases to new places where they devastate natural populations. Islands have been particularly hard hit.

The current extinction also differs because it is nearly all terrestrial. The five ancient mass extinctions were established based on the marine fossil record. Currently, extinction in the oceans due to human activity is virtually nil. It has only been recently that technology has allowed large scale exploitation of the marine environment and an accompanying dramatic increase in defaunation.[27] Anthropogenic stressors on the ocean, such as climate warming, and acidification are just beginning to have an impact. The fossil record suggests that these can become devastating if allowed to continue.

Related to this is what Anthony Barnosky and his colleagues have identified as "severe data comparison problems" between how paleontologists and biologists measure extinction.[28] Paleontologists see only what is preserved. Many of the modern animals becoming extinct are small and limited in their range, and of course with low numbers of individuals fewer may be preserved. In a paper I wrote in cooperation with Felisa Smith and Kate Lyons, we determined that very few of the mammal species threatened with extinction today are known as fossils.[29] Not only were these species likely to disappear, but their very existence will be unknown if written records vanish in the far future. The result is that the extinction will not look as severe when viewed from a time in the future.

A final key difference is scale. Although many of the recent extinctions are dramatic and disturbing, they do not yet reach anywhere near the level of the great mass extinctions of the past. At this point in time, the total extinction is about 1 percent, far less than the 75 percent characteristic of the P-Tr or K-Pg extinctions. However, the *rate* of current extinction is far above the usual background rate at times removed from mass extinctions based on the fossil record. Based on current estimates, 26 percent of all mammals and 41 percent of amphibians are threatened with extinction.[30] Although we have not reached the disastrous levels of past episodes, we might well get there over the next few centuries.[31]

The modern extinction (the Sixth Extinction) is far from being as severe as the great extinctions seen in the fossil record, so there is time to act to save what we still have. The aftermath of the previous extinctions teaches us that recovery of ecosystems takes tens to hundreds of thousand years. Diversity may take tens of millions of years to reach preextinction levels. If left alone, life will locate a path to recovery. However, at best, this process takes far longer than the span of modern human existence. Best to stop the loss now!

I have omitted one extinction. Humans certainly played a key role in the extinction of many large mammals at the end of the last ice age, the Late Quaternary Megafaunal extinction.[32] In North America, of nine herbivore species weighing more 1,000 kilograms (megaherbivores), such as mastodons and mammoths, all disappeared ten thousand years ago.[33] Of the nine megacarnivores (>100 kilograms), all but brown bears and jaguars vanished. There is a similar picture in South America, with the loss of all twenty megaherbivores and two of the four megacarnivores. In Australia, the extinction occurred earlier, about forty-five thousand years ago. The one megaherbivore went extinct, along with all but one of twenty-one herbivores in the 45 to 999 kilogram weight class.

The extent to which humans directly caused the extinctions, as opposed to the impact of rapidly changing climate at end of the last age, has been fiercely debated among paleontologists and archaeologists for decades. There are strong proponents for a primary role for climate, for the major player being human hunters, and for a fatal combination of both.[34] What is not in doubt is the impact of the extinction on communities and ecosystems that no longer had access to such huge animals.[35]

Whether they played a primary or a secondary role, extinction of the megafauna was the earliest clear signal we have of the human impact on the Earth.[36]

In the following millennia, this impact became more and more profound as humans spread throughout the world; agriculture changed patterns of land use, such as deforestation;[37] and the level of greenhouse gases in the atmosphere began to rise. The discovery of the New World by Europeans and the ensuing waves of colonization, immigration, displacement, and trade moved species, such as rats, around the world and produced extinctions of local species and the homogenization of the global biota.[38] The Industrial Revolution and the huge consumption of fossil fuel, which continues to this day, began to dramatically change the atmosphere. The level of carbon dioxide in the air in 2020 has reached levels not seen for millions of years. And, of course, the number of humans and their animals has attained unprecedented levels.[39] Overall, the level of human impact on Earth systems has become equal to or exceeds that of many natural processes. We are in an era of drastic and rapid global environmental change. This has led many scientists to push for the designation of a new episode of geological time, the Anthropocene.[40]

The best known of the many components of the Anthropocene are the changes in atmospheric and oceanic chemistry and temperature, usually summarized by the phrase "global warming." That the global climate is warming is, like evolution, a fact, not a theory. Nevertheless, like evolution, it is politically contentious. One of the recurrent themes of climate change denialists is that we can't blame humans because "climate is always changing!" My general response to this is a shrug and to say, "Aren't we the ones who told you that?" The idea that climates of the past have been different from those of today is obvious to any paleontologist or other student of Earth's past. Data from the fossil record is what first gave scientists an appreciation for changing global climates, and this data remains essential for unraveling the causes, patterns, and consequences of these shifts. It is our appreciation of these changes that makes us so worried; we know the historical context of our modern climate and that it is increasingly anomalous.

During my senior undergraduate year, I worked for James Hays at the Lamont-Doherty Geological Observatory. I became intimately familiar with a tiny fossil called *Cycladophora davisiana*. This organism is a radiolarian, one of a group of single-celled organisms that make their gorgeous shells out of silica. What makes *Cycladophora davisiana* special is that it likes cold water; in the far northern Pacific, it can make up almost half of the radiolarians. My work-study job was to take a sample of sediment from a core, prepare a slide covered with radiolarians, and count how many were *Cycladophora davisiana*. What stays with me to this day is that as I went farther down in the core, and thus further back in time, the number of specimens of the species increased and rapidly reached a peak. I was seeing firsthand, from a tiny fossil, direct evidence for the cold oceans of the last ice age and the warming since then.

Cycladophora davisiana is an example of a climate "proxy," an indicator of ancient climates before the time of instrumental or even historical records. There are a huge range of available climate proxies, covering nearly the entire range of Earth history. Most are directly or indirectly paleontological in origin. These include such crude indicators as ancient reefs, signifying warm tropical seas, or coal, formed under warm and wet conditions. Many proxies are far more precise.

Plants are highly sensitive to climate, so their fossils are key tools in determining ancient environments. Reconstructing plant communities based on fossil pollen and leaves have allowed detailed reconstructions of terrestrial climate throughout the Cenozoic. Fossil pollen has been a key tool in reconstructing climate change during and since the last ice age. For example, a core in Northern Illinois shows an abundance of spruce pollen after the glaciers receded. Spruce is currently rare in Illinois but is common far to the north in Canada. In recent years, spore data has been supplemented by a host of new isotopic methods, but still derived from fossil materials. Such data are a key archive of environments prior to and following human disturbance.

The form of leaves reflects their growing conditions, so they can be used to reconstruct temperature and rainfall.[41] The density of pores (stomata) on fossil leaves is inversely related to carbon dioxide levels in the atmosphere and is a major tool in interpreting the history of this greenhouse gas deep into geological time, providing key context for its modern rise.[42] This data helps confirm the close correlation between atmospheric carbon dioxide and global temperature and demonstrates that it has been at or below 300 parts per million (ppm) for the last twenty million years.

In the oceans, the shells of tiny single-shelled organisms, the calcareous foraminifera and the siliceous radiolarians ("forams" and "rads" to the cognoscenti), are microfossils that preserve the ocean conditions at the time they were formed. Some of this data comes from determining which species were present. The counting of *Cycladophora davisiana* that I performed was an example of this. The other key data comes from measuring the isotopic composition of the shells, which is a sensitive climate indicator (see chapter 6). My data collection was part of two major projects. The first of these used the data from microfossils to map the temperatures of the sea surface during the peak of the last ice age, about eighteen thousand years ago. The second used precise dating of the changes over time to demonstrate that the multiple advances and retreats of the ice sheets of the last three million years were controlled by small periodic changes in the Earth's orbit,[43] a hypothesis put forth decades before by Milutin Milanković. I am inordinately proud of my tiny place in this study.

The climate record from microfossils, more recently supported by the study of trapped gases from cores in the Antarctic and Greenland ice sheets, gives us the context for studying the climate impact of humans. Climate has gradually cooled for most of the last sixty-six million years, culminating in the onset of glaciation in Antarctica about thirty-five million years ago and extensive Northern Hemisphere ice sheets three million years ago.[44] Since then, we have alternated between glacial and interglacial episodes, with the last ice age ending about eleven thousand years ago. We are currently in a relatively warm interglacial period and have been so for all of human civilization. Maximum warmth occurred about six thousand years ago. Left to natural processes, we should be beginning the long slide into the next ice age, although this is many thousands of years away.

But we have changed the game. Confirming the data from plants, the record of carbon dioxide from the ice cores tells us that for at least the last eight hundred thousand years, carbon dioxide never exceeded 300 ppm. As of 2020, Co_2 levels exceed 400 ppm and are projected to increase to over 600 ppm and perhaps as high as 1,000 ppm by the year 2100, depending on future production. These levels have not been seen for fifty million years. And the rate at which we are approaching these levels and the associated warming is unprecedented. As with the case of the Sixth Extinction, paleontological data tells us that we are entering unknown territory.

What these numbers give us is context. Not long after the 2010 BP Deepwater Horizon explosion and oil spill, a recently retired geologist from BP spoke on our campus. When a student raised the issue of carbon dioxide emissions, he pointed out that "in the Ordovician, CO_2 levels reached 6,000 ppm!" My reaction was: "So what? There were no land plants or forests. High seas covered most of the continents. It was a different world and not relevant to what is happening today." Context!

Paleontologists are rallying to use their knowledge and methods to address the critical problems for humanity produced by global environmental change. This is best expressed in the relatively new field of "conservation paleobiology," which two of its founders (Karl Flessa and Greg Dietl) defined in 2011 as "the application of theories and analytical tools of paleontology to biodiversity conservation."[45] Conservation paleobiologists use the fossil record to understand the evolutionary and ecological responses of present-day species to changes in their environment, with the goal of guiding decisions on conserving and restoring modern biodiversity. Using the fossil record for this purpose provides many advantages: a longer time scale perspective on ecological change than is available solely from modern studies; a wider range of environmental conditions (e.g., ice ages) is explored than currently exist; and direct evidence of how and how rapidly species responded to past environmental changes are known.[46] Using the fossil record also allows us to sort out which extinctions are anthropogenic as opposed to natural and which species are native versus invasive.

Conservation paleobiology has a critical role in determining the baselines for ecological restoration. Restoration ecologists hope to repair ecosystems that have been damaged or even destroyed by human activities, and the state they want to restore them to is the "baseline." This can be controversial; for example, when restoring forests in Illinois, do we want to restore them to the state before European arrival or before native peoples migrated to the area and made their own alterations? In many cases, the baseline for restoration long predates our direct knowledge of the ecosystem. This is where paleontological and related data, "paleoarchives," come in.

For example, prior to construction of numerous dams on the Colorado River, its water freely reached the Gulf of California. Flessa and Dietl point out that no detailed scientific studies were made on the environment at the outlet prior to 1960, when the flow was shut down. Remarkably, there are huge shell beaches at the mouth of the river made up of trillions of shells. Dating these shell deposits show how incredibly productive an ecosystem once lived there. Restoring at least some of the flow annually could help restore this environment.

Oysters in Chesapeake Bay are both economically and ecologically valuable. They have been a human food source for thousands of years and greatly improve water quality. Unfortunately, the combined effects of overharvesting, pollution, dredging, and other environmental insults have reduced harvests by 98 percent since the late 1800s. As was the case in the Gulf of California, restoration is hampered by a lack of data on what the oyster reefs were like prior to human disturbance. Rowan Lockwood examined populations of Pleistocene oysters in the area and showed that those oysters grew larger and had longer life spans than oysters in the Chesapeake today, probably as the result of harvesting practices.[47] Abundances were also far higher and well beyond currently set restoration targets. The ability of the oyster population to filter water was an order of magnitude greater than today. As Lockwood says, "The fossil record contains a wealth of information on how oysters have responded to a variety of stressors in the past, which can help us predict how they will respond in the future." This information is key to restoration.

One of my courses in recent years was an undergraduate honors seminar on the Anthropocene and the Sixth Extinction, and we read and discussed Elizabeth Kolbert's book, *The Sixth Extinction: An Unnatural History*.[48] Nearly all of the students had heard about climate change, but few had ever heard about the other components of global environmental change. I was happy to raise consciousness on these topics, but it was very hard not to convey despair and anger over the fate of the Earth. Unfortunately, the recent history of action on climate change and conservation provides only limited grounds for optimism. We can only plan for a future of environmental disruptions, both fast and slow, if we understand what has happened before. Paleontologists, because of their intimate knowledge of the history of environmental catastrophes, should be active participants in this planning. We know who lived and who died, and we can tell their story.

9

LESSONS FOR AND FROM THE LIVING

My doctoral thesis documented the fossil record of the eurypterids, an extinct arthropod group related to living horseshoe crabs and scorpions. Like those animals, and unlike other marine arthropods such as crabs and lobsters, eurypterid external skeletons lacked hard minerals and thus are rarely preserved. One of my doctoral committee members, the late Tom Schopf, asked me a perceptive question: "If they don't preserve well, how can you trust what you know about their fossil record?" I had no ready answer. After I completed my thesis, I decided I would make this problem a focus of my research.

The following summer found me and my new (and very accommodating) wife at Friday Harbor Laboratory (FHL) on San Juan Island, Washington. I had been there two years previously to take summer courses, and I knew it was the ideal place to study why marine arthropods do or do not become fossils. The waters around FHL teem with a wide variety of arthropods, there are a diversity of marine environments, the laboratory is well equipped, there are many other scientists to talk to, and it is one of the most beautiful places I have ever been.

The scientific question I was asking was a simple one: What happens to arthropods after they die? Taphonomy is the study of fossil preservation, and I was examining the earliest stages of the process. My test animal was one of the very abundant shrimps, easily caught off the dock at the lab. After killing them by freezing (this is when I learned the biologist's euphemism "sacrificing"), I carried out two sets of experiments. The first of these was to bury the shrimp in the mud or sand at low tide and return the next day to see if they were still there. The answer was usually no; at best I would find a tiny fragment. My guess was that various scavengers, such as crabs, had gotten to them. In the other set of experiments, I placed the shrimp in jars with seawater and watched what happened to the carcasses over time.[1] After a few days, the stink became intense. My lab mates informed me that I would have to go out on the dock, well away from the lab building, to open my jars. Despite the stench, I was able to make some important observations on how shrimp fell to pieces over time. Primitive as this research was, it stands as one of the earliest attempts to study fossil preservation experimentally.[2] Others have done far more sophisticated versions in recent years.[3]

This type of study illustrates four important aspects of paleontological research. First, the study of preservation; the where, how, and when dead things are preserved or not preserved is essential for understanding the fossil record. Second, although perceived as a purely historical science, experiments are integral to many areas of paleontology. Third, paleontologists study living organisms and environments to interpret ancient ones correctly. Finally, we insist that truly understanding the modern biological world requires knowing its history. In this chapter, I explore the interactions between the living and the long dead and show that paleontology is an integral and inseparable part of the life sciences. Paleobiology is as much a biological science as is cell biology or genetics. These interactions can be roughly divided into "bringing the dead to life," i.e., reconstructing the biology of long extinct organisms, and "the past of the present," integrating the fossil record and the modern biological world to understand the history of life on Earth. As Douglas Erwin (Smithsonian Institution) put it in 2017: "Fossil material is embedded in a geological context, but in studying fossils we often pursue biological questions."[4]

Understanding the processes of fossilization is the critical first step to unraveling the biology of ancient organisms. We need to know what is there, what is missing, what has been changed, and why. Fossilization is akin to signal processing; there is an original biological signal, which is composed of myriad anatomical, physiological, ecological, and behavioral properties that we use to describe living organisms. Then there is the altered signal studied by paleontologists, which might be of high fidelity and cleanly record the original biological signal or low fidelity with a highly degraded or modified signal. Understanding how the signal is modified aids in reconstructing the original information, such as providing evidence about the environments in which the fossils formed. Every step in fossilization is not simply loss of information; instead, each phase adds critical evidence: How did the organism die? Who scavenged it? Which bacterial decay processes were operating? What was the environment it was buried in? How much time does the fossil deposit represent? etc. Taphonomy is not, as it is often misconstrued, the study of how the fossil record is biased but is a much broader consideration of how preservation affects the information contained in the fossil record. Studying preservation also reveals that fossilization has changed over geological time, providing critical clues about the evolution of the entire Earth system. Rather than being viewed only as a bias or distortion of the biological signal, taphonomic processes have their own signal that we use to aid our understanding.

The recognition that fossilization is not simply "bias" is part of a welcome attitudinal change among my fellow paleontologists.[5] We no longer shuffle our feet and apologize for what Darwin famously called "the imperfection of the fossil record," which led to the impression that the fossil record is hopelessly incomplete. That is very far from the truth. We now affirm that the fossil record contains a great deal of information, much of which is unique to it.

Paleontologists study all phases of fossilization, from the original biological signal to the eventual recovery and description of fossil remains. This research is often predictive and frequently based on studies in modern environments. The first modification of the biological signal occurs when an organism dies or sheds parts of itself (for example, shark teeth, arthropod molts, or tree leaves) to produce nonliving remains. Like forensic scientists, we hope to unravel the cause of death. Sometimes the culprit is obvious. Some twelve million years ago, a massive volcanic eruption in Idaho produced an ash plume that buried herds of ancient rhinoceros and other animals far away in Nebraska.[6] The animals died directly by being buried under the ash or from the effects of breathing in the ash. You can see their remains and their continuing excavation at the Ashfall Fossil Beds State Historical Park in eastern Nebraska.

Being eaten is a less spectacular but far more common way to die. Large lakes teaming with fish and other life forms existed in Wyoming, Utah, and

Colorado from about fifty-five million to forty-five million years ago, form-ing the spectacular Green River fossil beds. Among the huge number of fish fossils from this site are many that perished in the act of eating another fish, sometimes even a member of their own species.[7] If you have been to any ocean beach, you may have noticed that many shells have distinctive conical holes in them. These are drill holes made by voracious predatory snails, usu-ally a species of moon snail (naticids). Once the snail has penetrated the shell, it can rasp and suck out the flesh of its living prey. These distinctive bore holes have been found deep in the fossil record, at least as far back as the Cretaceous or possibly older. The distinctive shape of snail borings makes them a favor-ite subject of paleontologists, such as Patricia Kelley and Mike Kowalewski, who are interested in documenting the history of predation over time. It is an ecological interaction documented over tens of millions of years.[8] The only comparable data is the feeding of herbivorous insects on plants, which leaves distinct feeding marks on leaves and wood,[9] providing a record of herbivory that goes deep into Earth history.[10]

Of course, there are many other ways to die (disease, starvation, etc.), but the result is always a dead thing. Numerous indignities can befall the unburied dead. Scavengers can consume and tear them apart; waves or rivers can move them; a passing herd can step on them. And, as demonstrated by my shrimp studies, organic materials rot. Decomposition by microbes is a fundamental biological process that is essential to recycling nutrients. Without it, we would find ourselves buried in dead mice and cockroaches and starving to death. As bodies rot, they fall apart. Shells of bivalved animals separate; bones and teeth come apart, disarticulating the skeleton into separate pieces; branches break off dead tree trunks. The individual pieces can be further damaged by a range of physical, chemical, and biological processes. The preponderance of fossils are shells, bones, teeth, and wood—the parts of organisms that are most resistant to decay.

Decay, decomposition, and disarticulation must be studied in the modern world. In a remarkable piece of research, Anne Kay Behrensmeyer has been carefully monitoring the fate of vertebrate remains in Amboseli National Park in Kenya for more than forty years.[11] The insights she has gained allow her to interpret the formation of ancient vertebrate fossils in places such as the Siwalik mountains of Pakistan; they also provide critical information on the environments of early humans.

Behrensmeyer's approach to taphonomy has become a touchstone within paleontology and has inspired many other studies. In the early 1990s, a disparate group of paleontologists decided to examine long-term processes of breakdown on the continental shelf. Their project, the Shelf Slope Experimental Taphonomy Initiative (SSETI), used submersibles to place a variety of shells and wood on a range of locations on the seafloor of the Gulf of Mexico in 1993, and for more than a decade they tracked what happened to them.[12]

Given the plethora of processes that decompose living things, the vast majority of remains never make it to become potential fossils. Those that do must, in some manner, be removed from the "slings and arrows" of decay, either before or after burial. There are many ways for this to happen. The water above the seafloor may lack oxygen, inhibiting scavenging and many forms of bacterial decomposition. The remains may be rapidly buried, again protecting them from many of the agents of decay. Or they may be trapped in unusual settings such as amber or tar. When the conditions are right, we can get thick accumulations of shells or bones.[13] In some remarkable cases, we can preserve information about the soft tissues of organisms, sometimes including the original organic molecules. In 1970 Dolf Seilacher named these sites Konservat-Lagerstätten (roughly translated as "conservation mother lodes"; usually shortened to just Lagerstätten), where unusual amounts of biological information are preserved. Lagerstätten provide windows into what is rarely preserved and are the subjects of intense scientific interest.

The most celebrated Konservat-Lagerstätte is the Burgess Shale in the Middle Cambrian of British Columbia, made famous by Stephen Jay Gould's 1989 book *Wonderful Life*. Discovered more than a century ago, the Burgess Shale preserves a huge array of soft-bodied organisms, many of them strange to the modern eye, as well as imprints of the soft tissues of mineralized organisms. In recent decades, numerous additional Cambrian sites of similar preservation have been discovered, such as the older Chengjiang site in China. The detail in these sites is often exquisite, down to the structure of arthropod brains and the central nervous system.[14] This tells us that such structures, and the behaviors they made possible, originated more than five hundred million years ago. Cambrian Konservat-Lagerstätten are critical clues to understanding the early days of animal life on Earth.

Although there are many other Konservat-Lagerstätte in Earth history, the type of preservation seen at the Burgess Shale and Chengjiang sites, labeled

"Burgess Shale Type" (BST), seems to disappear sometime in the Early Ordo-vician. The window of time during which BST deposits occur reflects an exceptional combination of biological and environmental factors during that period of Earth history. Other periods are likewise characterized by condi-tions unique to them, which is reflected in the nature of fossil preservation. Taphonomy has become part of the larger effort to understand the evolution of the Earth system over time.

A common misconception is that making a fossil requires that the remains become "petrified," that is, turned to stone. Although this is often the case, as with petrified wood, it is not necessary. Fossilization is *any* pathway that pro-duces a fossil that a paleontologist will study; it does not require modification of the original material.[15] Among the most exciting fossils are those in which the organic molecules are themselves preserved. Recent years have seen a tre-mendous growth in "molecular paleontology," which attempts to identify the molecules that are present and the information they can give us about the original organisms. Much of this has been made possible by the development of new, highly sensitive methods to detect the presence of organic molecules. Preserved organic molecules range from carbohydrates (cellulose, chitin), proteins and lipids, to nucleic acids.[16] Unfortunately, the Jurassic Park dream of intact dinosaur DNA is just that, a dream. DNA degrades too quickly to be preserved intact. New techniques have, however, allowed much of the DNA of more recently extinct organisms, such as mammoths and our cousins the Neanderthals, to be reconstructed and compared with living animals.

In 1994, Lowell Dingus published the book *What Color Is That Dinosaur?* His answer was "any color you want." That is no longer the case. One of the most exciting recent developments is recognizing that many of the molecules, such as melanin, which is responsible for color, have survived in fossil organ-isms.[17] Not only did many dinosaurs have feathers, but the feathers also had distinct color patterns. This suggests that, like their surviving descendants the birds, dinosaurs carried out sexual displays. And some dinosaur eggs were blue green![18]

The animals of the Burgess Shale and Chengjiang, or for that matter any fossil, would make little sense if we were not familiar with the breadth and organization of modern life. We use this knowledge as a guide to resurrect the biology of extinct forms. In my introductory paleontology class, the very first lab is the dissection of the common clam (the quahog, *Mercenaria*

mercenaria). Clams are bivalves, and the animal is contained within two hard shells. The shells meet at a hinge; on one side of the hinge is a pair of muscles used to close the shell and on the other side is a hard rubber ligament that opens the shell.[19] Both the muscles and the ligament create impressions on the surface of the shell. They also act to hold the two valves together. Clams live buried in the sediment and use a tube (siphon) to reach the water at the surface that they pump to feed, and the siphon also leaves a distinct notch mark on the inside of the shell that is directly related to its size: the larger the notches, the bigger the siphon.

After examining the clam, I give my students the separated shells of another species. They locate where the muscles were based on the distinctive scars they left behind and notice the very deep notch produced by the siphon. They also find the hinge and the former location of the ligament. Using the dissected clam as a model, they reconstruct the soft parts and realize that these animals had a huge siphon and thus must live deep in the sediment. I then show them a preserved specimen where the soft tissues are still present, and they can see that they are correct.

This simple example illustrates the basic steps of reconstructing a fossil organism. First, you must know the anatomy and ecology of living organisms. As a student, I took a course in marine invertebrate zoology; my colleagues in vertebrate paleontology studied (and now often teach) comparative anatomy.[20] You learn to recognize the distinctive marks, such as muscle attachments, that soft tissues make on hard parts. Underlying this is a fundamental concept of evolution—similarities in anatomy among organisms exist because they are all related, the concept of homology. We can assume that the paired elliptical scars inside a Jurassic bivalve shell are attachments for muscles just as they are in recent forms. Using our knowledge of the anatomy of modern forms guides us in reconstructing the missing soft tissues of their ancient relatives. The same basic methods can be applied to reconstruct a Cretaceous dinosaur, a Miocene bat, or a Pennsylvanian tree.

A reconstructed fossil organism is still static. To make it live again, we need to know what its various parts did in life, that is, their function. This is the goal of functional morphology, which attempts to determine how ancient organisms ate, moved, mated, etc.[21] One approach is to recognize the close relationship between the form of an organism and what that form does in life. For example, Steven M. Stanley (University of Hawaii at Manoa) published a seminal study on how the shapes of bivalve shells, such as clams, oysters, and scallops, were closely correlated to their life habits.[22] Using this knowledge, one can pick up a fossil bivalve, reconstruct its soft tissues, and know where it lived in the ocean and how it ate.[23]

This approach, however, has limited value when working with extinct animals. There are no living pterosaurs, so how could we determine whether their wings were used for flying or gliding? Or, if they were flyers, did they fly like albatrosses, falcons, or pigeons? We can look at a pterosaur wing and see which bird wing it matches best, but this just suggests a hypothesis. To test a hypothesis like this, the best thing to do is utilize the principles of biomechanics, the science that applies the laws of physics and the methods of engineering to understand function in living beings, to understand fossils. Paleobiomechanics is another field in which paleontologists can conduct experiments.

The best course I ever took was in biomechanics, taught at Friday Harbor Marine Laboratory by the late Steve Vogel (Duke University)[24] and Michael LaBarbera (University of Chicago). From them I learned how to perform hands-on experiments using physical models to test ideas about how animals lived in the past. One group to which I have applied this approach are crinoids. Crinoids are a strange, wonderful, and beautiful group of animals. Strange because at first glance many of them look more like plants than like animals; their common name of sea lily reflects their plantlike appearance. Wonderful because they are among the most common animal known from Paleozoic oceans, and they show a fantastic diversity of form. Beautiful because, frankly, they are one of the most visually appealing groups of fossil animals. One common crinoid from the Silurian is *Ancyrocrinus*, the "anchor crinoid," which lived on soft muds. As the name implies, the end of the animal that held it in place on the seafloor (the holdfast) looks very much like an anchor, with four prongs projecting upward from a round bottom. But what kind of anchor? Previous papers suggested that the prongs were sharp at the tip and dug into the sediment. This turned out not to be true; the prongs were blunt. When I dragged a model of the holdfast across the mud, the prongs did not dig in. Instead, working with my then undergraduate student Jennifer Bauer, we experimentally showed that the holdfast was designed to be buried in the sediment. What was exciting was that it could bury itself in a current.[25] I submitted this to the university as a possible patentable invention as a boat anchor; they agreed it was novel but were not willing to pay for the patent (anyone interested?).

My somewhat crude experimental models were laboriously constructed of metal, plastic, and plaster. Nowadays I can easily print accurate 3-D models. I could replace the flow tanks, strain gauges, and force meters that I was trained to use with simulated physics in a virtual environment. The advent of engineering methods such as finite element modeling and computational fluid mechanics have led to the advent of a "virtual paleontology," allowing increasingly sophisticated interpretations of the biology of extinct organisms.[26] To determine the function of a pterosaur wing, we could either build a model of one and put it in a wind tunnel or put a virtual model in a simulated airflow.

At the vanguard of this approach is Yale postdoctoral student Elizabeth Clark. Brittle stars, close relatives of starfish, move by lashing their thin arms. They have about one thousand tiny moving parts but no central brain to control their movement. Liz makes biomechanical models of fossil brittle stars and studies them using methods adopted from mechanical engineering. She has made 3-D scans of superbly preserved specimens from a Devonian Lagerstätte and produced 3-D printed models that she can hold and manipulate in her hands. Liz can also create virtual 3-D models that enable her to examine the range of movements at the joints in the arms. Her work has direct implications for robotics in understanding how movement can occur with decentralized controls.

As brittle stars move across the seafloor, they leave a trail behind them. Throughout the history of life, animals have walked, crawled, or burrowed through various substrates and left behind records of their passage. These preserved remains, not of the body of the animals but of their activities, are known as trace fossils (technically, ichnofossils). They include tracks, trails, burrows, borings, drill holes, bite marks, and many other ways an animal can signal to the far future that "I was here and did this!" Trace fossils are the best direct evidence we have for animal behavior in the deep past.

The key conundrum of ichnology, the study of trace fossils, is that it is often impossible to know who was there and what they were doing. Many kinds of organisms and many types of behaviors produce similar traces. One approach to unraveling this is to study the behavior and movements of living animals, an approach known as neoichnology. For example, Dan Hembree (Ohio University) has intensively studied the burrows made by a variety of desert dwelling animals, such as scorpions, to see if they could be distinguished in the ancient record. The Emory University paleontologist Anthony Martin has written a lively and entertaining book about the traces that life leaves behind on Georgia coastal islands, a place he has studied for many years.[27]

Ichnology is yet another area of paleontology amenable to experimental investigations, both in the computer and with real animals. Hembree has varied the nature of the soils to see if they changed the shape of the burrows. I wrote computer models for trace fossils that suggested that much of the variation in the form of preserved trails was due not to differences in behavior but to the distribution of resources. For her thesis, my PhD student Karen Koy (now at Missouri Western State University) successfully tested this idea using modern snails and nematode worms.

Swimming is another animal activity with ancient origins, and it is the focus of Christopher Whalen, a PhD candidate at Yale University. He is the somewhat rare exception, a paleontologist who does not choose between vertebrates and invertebrates, a division he sees as a problem: "Why not look at all groups?" Chris is interested in the evolution of actively swimming organisms in the middle of the Paleozoic era (roughly 540–360 million years ago). Key to this research is reconstructing the evolutionary relationships within two very different groups: the nautiloids, early shelled ancestors of living cephalopods such as octopus and *Nautilus*, and the early jawed vertebrates, ancestors of nearly all living vertebrate groups. This attention to relationships is an example of "tree-thinking."

What is tree-thinking? The concept of homology, the idea that many of the similarities among organisms are due to their being related, underlies phylogenetics, the determination of how organisms are related to each other. Phylogenetics depends on three fundamental ideas: that every living thing on Earth is related by descent in some way to every other living thing; that many features of their biology are shared because of that; and that novel features appear over evolutionary time and are passed on to the descendants of the first organism to have them. The last of these implies that the more "novel things" (technically, synapomorphies) two organisms share the more closely they are related. These relationships are usually expressed in the form of a branching diagram, called a cladogram, that resembles a tree. The methods for constructing a tree are known as cladistics.[28] Closely related organisms are on adjacent branches and share more novel features, such as the lack of a tail in apes and humans; organisms on distant branches may share only a few things in common, such as the five fingers of the hand of humans and frogs. Many of the fundamental questions in biology and paleontology can only be addressed by knowing the shape of the tree of life—by "tree-thinking." Bhart-Anjan Bhullar (box 9.1) insists that only by tree-thinking can we ask the right questions about the evolution of life.

BOX 9.1

Bhart-Anjan Bhullar is an assistant professor at Yale University. From an early age he read both literature and natural history. Growing up in relatively rural places in the Midwest, he spent all his time outdoors, which gave him both a sense of independence and of collaboration with other children. Anjan notes that his high school teachers were very influential, such as the AP biology teacher who took them out "herping" (looking for reptiles and amphibians). When he entered Yale as an undergraduate, he had a major in molecular biology. He came, however, to dislike the "ahistorical" aspect of that field; that is, the research was divorced from "how did this come to be?" In his sophomore year, Anjan by chance met Jacques Gauthier and ended up taking his course on lizard evolution. He went on to the University of Texas at Austin for a master's degree and was inspired by Tim Rowe from whom learned about CT scanning, a technique for making studies of three-dimensional anatomy in fossils. At Harvard Anjan studied for his PhD with Arkhat Abzhanov, who is not a paleontologist but is a specialist in evolutionary developmental biology, a field with tremendous potential for integrating with paleontology. Anjan characterizes himself as a biologist who "thinks in deep time." And admits he always wanted to be a paleontologist.

Making such trees is not simple. There are many millions of species on Earth and many millions more that are extinct, and there are an uncountable number of ways they all can be related. Determining the most likely pattern of relationship can be both conceptually and technically incredibly difficult because many different trees may be compatible with the available data. Phylogenetics is one area that has been tremendously transformed by the advent of powerful computers and new methods of molecular analysis. It is also an area in which paleontologists and biologists often work together yet are sometimes at odds.

For decades, determining the relationships among organisms, both living and extinct, was primarily based on morphology. Although some information may be missing, data on both modern and fossil forms could be combined into the same analysis. The development of new molecular methods, such as the rapid sequencing of DNA, has led to an explosion of new analyses. Although the trees developed from these techniques can match those produced by studying morphology, for a range of biological and technical reasons they often produce contradictory results. One of the most important of these is that molecular methods can use only living organisms, which represent a fraction of the entire tree; that is, only those branches that have survived until today. As Dave Raup once said, "to a first approximation, all species are extinct."

Charles Marshall (University of California, Berkeley) put it forcefully in a 2017 commentary:

> In the absence of a fossil record we would not know that the first tetrapods were not fully terrestrial, or that they had more than five digits per limb, that the first lungfish were marine, that fivefold symmetry is not ancestral for echinoderms, that many of the synapomorphies of living birds evolved before flight, nor could we have inferred the unexpected morphology of *Ardipithecus*, which lies close to the last common ancestor of humans and chimpanzees.[29]

Similarly, Mark Pagel wrote in *Nature* in 2020:

> Without fossils, all evolutionary scientists, whether studying speciation and extinction or attempting to reconstruct the features of distant ancestors, need to be aware that the evolutionary processes they identify are those that operated in the species that would survive and eventually leave descendants in the present. We can't be sure what was going on in those that went extinct. It is the evolutionary version of the observation that history is written by the victors.[30]

The Yale vertebrate paleontologist and herpetologist Jacques Gauthier would agree. His 1984 thesis was the first attempt, using the then novel method of cladistics, to produce a tree that included all reptiles, living and fossil. Jacques considers it his job to find the novel features that join branches of the tree together and insists that you "need fossils to get the right tree! Modern forms alone don't work."

Trees of relationship also imply timing of evolutionary events. Each fork in the tree represents the ancestors of two groups of species splitting from each other. The closer two species are on the tree, the more recently they diverged from a common ancestor. For example, if we have three closely related species at the outermost "twigs" (tips) of the tree, they represent three splitting events: (1) when the three split off from their common ancestor, (2) when two of them divided off from the third, and (3) when those two split from each other. We can use this concept to count how many splits separate any pair of species. But how long ago did these splits occur? For example, when did all the placental mammal groups, from whales to bats, diverge from each other? The fossil record strongly suggests that this happened soon after the end-Cretaceous extinction of the dinosaurs.

Molecular biologists have tried to get at this answer by assuming that the rate of genetic change caused by mutations following a split is more or less constant, the so-called molecular clock. If we know this rate of change, we can use it to estimate how long ago the ancestors of two living species diverged. To calibrate this clock, they use the fossil record of these species or their ancestors. The result is a tree of life based on molecular biology that includes a time dimension.

The long-standing issue, just now beginning to be resolved, is that the ages given by the molecular clock are almost always older than the fossil record indicates, sometimes by quite a lot. In the case of the divergence of placental mammal groups, some clock estimates put it back into the Cretaceous, long before the fossil record says it occurred. The same problem occurs with determining the dates for the origin of animals.[31] The discrepancy could be due to lack of preservation in the fossil record, incorrect assumptions or data underlying the use of the clock, or some combination of both.

Development—the study of how an organism goes from the fertilized egg or seed to the adult—is one area with a rich and fruitful exchange between paleontologists and biologists. This interaction has been beautifully described in Neil

Shubin's book, *Your Inner Fish*. As expressed to me by Sarah Tweedt (box 9.2): "all multicellular organisms develop from a single cell to an adult. The big question is: How did this process begin, and how did it evolve? The origin of animals is the origin of development; the origin of animal groupings such as phyla is the origin of development." Sarah, trained as a biologist, sees herself as a "bridge and translator between molecular biology and paleontology."

BOX 9.2

*Sarah Tweedt, currently a postdoc at the University of Colorado, did not take any courses in paleontology until she was a graduate student at the University of Maryland where paleontology courses are taught by NMNH curators. Sara grew up in the Tidewater area of Virginia, near Chesapeake Bay. A childhood enthusiasm for biology led her to a bachelor's degree at the University of Virginia with a focus on molecular biology. Following graduation, she worked for more than two years as a technician in a cellular biology laboratory and as a bartender. During her vacations, Sarah crashed scientific conferences. At one of them, she heard David Jablonski talk about the relationships between developmental biology and paleontology. For someone who thought of paleontology as dinosaurs, this was a "whoa" moment. She soon looked for information on the two subjects online and learned about the cutting edge work of Douglas Erwin at the NMNH in this area. She became Erwin's doctoral student through a program at the University of Maryland that led to a coauthored paper in Science on the Cambrian explosion.**

* D. H. Erwin, M. Laflamme, S. M. Tweedt, E. A. Sperling, D. Pisani, and K. J. Peterson, "The Cambrian Conundrum: Early Divergence and Later Ecological Success in the Early History of Animals," *Science* 334, no. 6059 (2011): 1091–97.

All animals and plants, whether living or extinct, grow over their lifetimes. As they do, they often change shape, a process called allometric growth (think of how big the heads of human babies are!). These changes in shape and their timing relative to sexual maturity have long been invoked as key to understanding the processes of evolution. Steve Gould devoted most of his career to understanding this, as summarized in his highly readable 1977 book *Ontogeny and Phylogeny*. Although our bodies do not preserve what we looked like as a child, many other organisms do. Shelled organisms such as clams and snails have distinct growth lines that record their development. Many trees preserve tree rings that reflect annual growth cycles and year-to-year changes in climate. Remarkably, many dinosaur limb bones also show such banding, known as lines of arrested growth (LAGs). LAGs are generally interpreted as representing annual cycles of growth, and they are used to estimate how fast a dinosaur grew and how long it lived. From LAGs, it is estimated that the *T. rex* SUE at the Field Museum lived for twenty-eight years and grew very fast for the first twenty years of life.[32]

The relationships between the two histories—the life history of organisms and the evolutionary history of life—has been a subject of interest for well over a century. What has changed is our ability to track the nature and expression of genes during the process of development. Biologists can see where and when genes are turned on and off in the growing organism and which other genes control this process. What has been astonishing are the fundamental similarities of these genes across the animal kingdom. The set of genes that determine front-to-back in a fruit fly do the same job in a mouse. The genes in the mouse that say "make an eye here!" can be transplanted to make an eye in an eyeless mutant fruit fly. Beginning in the late 1990s, these discoveries led to the growth of "evolutionary developmental biology," usually shortened to evo-devo, which examines the genetic mechanisms underlying development and compares them among different groups of organisms. As discussed in chapter 8, some of the major questions in paleontology revolve around the issues of when major groups of organisms appeared, such as during the Cambrian Radiation. These are patterns of macroevolution. Again, as Doug Erwin says, "evo-devo has returned the focus to whether there are also distinctive *origin* accounts of macroevolutionary patterns." In other words, as the work of Shubin, Erwin, Tweedt, and many others make clear, studying development shines fresh light on many of the key questions we have about the history of life. At the same time, paleontology provides the timing and context for when key innovations in development occurred.

In an oft-cited phrase, Theodosius Dobzhansky said, "Nothing in biology makes sense except in light of evolution."[33] The fossil record provides the key evidence that life has indeed evolved. Without fossils, we would have a very biased perspective on life on Earth today. For example, by far the most diverse group of modern organisms are the insects. They have had a very long and complex evolutionary history. Dena Smith points out that insects are integral to our daily lives, so it is critical to know "How did we get here?"; that is, given the "great biodiversity of insects, how did they survive and diversify?"

Although evolution is the unifying concept in biological sciences, there is little exposure of today's biology students to the history of life as represented by the fossil record. Biologically trained paleobiologists, such as Sarah Tweedt and Bhart-Anjan Bhullar, were struck by the ahistorical nature of their education. Biologists unaware of the information in the fossil record will not consider it in their research. Biology without paleontology is incomplete.

III

EXPLORERS
OF DEEP TIME

10

THE EDUCATION OF A PALEONTOLOGIST

On *Star Trek: Voyager*, Commander Chakotay proudly states that he had wanted to be a paleontologist since he was six years old. When I was six, I wanted to be an astronomer. I spent a lot of time looking through telescopes, although the light-polluted skies of the Bronx provided limited opportunities for star gazing. To compensate, I made my indulgent dad, an English teacher, drag me to the Hayden Planetarium in Manhattan and to meetings of an amateur astronomy club in Queens. I read every basic astronomy book my public library had and even audited a college class on the subject. My dream died during my senior year in high school. A friend handed me a book called *Introduction to Astrophysics*. Given the title and that I had just finished a semester of AP calculus, I thought I could handle it. Wrong. The first page was nothing but equations. I realized that astronomy is astrophysics is physics is math. And to be a really good astronomer, one needed not only to be able to "do math" but to visualize the universe as a mathematical object. I decided this was beyond my abilities, so I started to look at other options.

Two occurred to me just as I entered Columbia. The first was geology. Plate tectonics had just swept the earth sciences. I attended a lecture about a time 5.5 million years ago when the

Mediterranean had dried up and was refilled by one of the biggest waterfalls in Earth history. This was exciting. The second was marine biology; I visited the aquarium and fell in love with the white whales. I ended up deciding to become a biology major. Unfortunately, the introductory biology class focused exclusively on cells and molecular biology; not a whale (or any other beast) in sight. I quickly switched to geology.

At the same time, I moved to campus and got a work-study job at a blood bank to help pay for it. Because a mistake at a blood bank could be fatal, they gave me very little actual work to do. Luck and parental guidance now stepped in. I read an article in *Natural History* magazine titled "A Trilobite Odyssey" by Niles Eldredge at the American Museum of Natural History. I told my mom that I found it fascinating. She suggested that I "call him and see if he has a job for you." So I did, and he did! An upperclassman (David Jablonski, now a renowned paleontologist at the University of Chicago) had just become ineligible for a work-study job and I was able to walk right into his slot.

Niles was the ideal person for me to work with. He had recently published, with Stephen Jay Gould, the concept of punctuated equilibria to describe the evolution of species. This ended up being one of most important paleontology papers of the late twentieth century. My original copy of this paper is a prized possession. He was also instrumental in introducing cladistics to palaeontology, a new method for understanding the relationships among organisms. Niles was at the cutting edge of the science. Just as important, instead of giving me make-work during my hours at the museum, Niles decided to train me in what I needed to know to do research in paleontology. This included learning how to understand and describe fossil material. By my junior year, he trusted my abilities enough to make me coauthor on a scientific paper, my very first. I now pay this forward in my own career, involving students in my research at every opportunity and making them coauthors.

When I was ready to go to graduate school, Niles urged me to go to the University of Rochester. The Rochester faculty included David Raup, one of the most important figures in the field, and rising stars (and new Harvard PhDs) Daniel C. Fisher and Jack Sepkoski. To go to Rochester, I had to turn down the opportunity to work with Steve Gould at Harvard. Needless to say, this shocked my friends. (Harvard had also offered me far less money. After paying for a private college education, I was done asking my parents to support me.)

Rochester at that time was an amazing place. In addition to the stellar faculty, my fellow graduate students were among the best in the country. Most would go on to highly productive careers in paleontology, and we are all still friends. We also had some stellar undergraduates, including Mary Droser and Arnold Miller, who became prominent paleontologists. Although I understood a lot of theory, I learned that I really knew very little about fossils! During my second year Raup recruited Curt Teichert, who had just retired from the University of Kansas. Teichert was an old-school paleontologist with a deep and broad knowledge of the fossil record. Once in class, he embarrassed me when I confused two well-known fossil groups: "They are all the same to you, huh?" It was a lesson that stuck; in my research and teaching, I try to emphasize both theoretical concepts and the fossils that underlie them. I also learned to deal with failure; my original thesis project with Dan Fisher foundered due to my own limitations. Thankfully, I was able to quickly devise and complete a new project with Raup.

Unfortunately, the program at Rochester collapsed. Sepkoski left, then Raup, and eventually Fisher. I followed Sepkoski to the University of Chicago, where I did my PhD on eurypterids, otherwise known as sea scorpions. Jack was more than willing to let me choose rather than dictate my thesis topic. Again, in addition to the excellent faculty, the best thing about Chicago was my fellow graduate students, many of whom remain lifelong friends.

I was very fortunate in my career. Working with Eldredge while I was at Columbia eased my path to graduate school at Rochester. Being Raup's and Sepkoski's student gave me a leg up on getting into Chicago. And being a graduate student at Chicago made me a much more attractive candidate when the time came to look for a job. I would like to think that my innate abilities (and my splendid good looks) are why I was successful, but coming from the right places and having the right connections was also a key factor.

When I started to interview my colleagues for this book, I thought my story of a delayed entry to the field would be unusual. I assumed that there would be two typical tales of why a paleontologist decided to study fossils for a living. The first was, "I loved dinosaurs ever since I was a kid, and I could not imagine doing anything else." The second was, "I grew up playing in places where fossils were all over the ground. I fell in love with them right then and there." This latter group would come from upstate New York or the Cincinnati region, where streams and roadcuts expose tons of easy-to-collect fossils.

I gradually learned, however, that my path was not unique. Although many were indeed paleontologists since the age of six, others came into the field as late as or even later than I did. In this chapter and throughout the book, I introduce some of my fellow paleontologists and tell the stories, often in their own words, of how they fell in love with fossils. I also share some biased advice on "how to become a paleontologist."

Among my colleagues are those who fit my mental image of the child fossil collector/dinosaur nut turned professional paleontologist. One of them is Lucy Edwards, a micropaleontologist at the United States Geological Survey who got her PhD from the University of California at Riverside:

> I was born in 1952, the middle of three kids, to parents who taught at the Medical College of Virginia and lived on a farm. I wanted to be a paleontologist since I got my first dinosaur model at age six and read Roy Chapman Andrews' books. In third grade, if you had asked me if I could go anywhere in the world I wanted, I would have said "Mongolia" (to hunt dinosaur eggs). By junior high, I figured I'd sunburn too badly, and maybe I wouldn't be a paleontologist. Every year or so, we'd visit the Smithsonian, and the dinosaur hall would be under construction. Then in high school the exhibit was finished, and when I saw it I was still hooked. I graduated from high school in 1969. I moved as far away from home as I could and went to the University of Oregon. By the end of my first year I knew I wanted to be a micropaleontologist.

David Jablonski, whose change in financial status opened the door for me at the American Museum of Natural History, knew he wanted to be a paleontologist by the time he was five years old.[1] A member of the National Academy of Sciences, Dave (or DJ, as he is often known) grew up near the American Museum of Natural History (AMNH), which he describes as his "home away from home," and was fascinated with the museum's magnificent dinosaurs. Dave, as is the case with many of us, had parents who encouraged him and gave him the freedom to pursue his "intellectually freewheeling existence." After entering Columbia University as an undergraduate, he received the work-study job at the AMNH that was later to be mine. This gave him the unparalleled opportunity to work with some of the top scientists in the field at a moment it was undergoing a profound change in its concepts and methods, a period that has been dubbed the "paleobiological revolution." He was also able to sit in on graduate seminars. "That's when I began to get a much better sense of paleontology as a profession and as a science." By the time he entered Yale University as a graduate student, Dave was already well on his way to becoming an established member of the field.

Jessica Theodor (University of Calgary) found her first fossil in her own backyard in Toronto. She took it to the Royal Ontario Museum (ROM), where it was identified as a crinoid (sea lily). Jess and her parents then discovered there were classes on fossils for kids on the weekends. The coolest thing is that they could wander through the halls of the museum. By the time she entered college at the University of Toronto, Jess knew she wanted to be a paleontologist. Fortunately, the university had an undergraduate paleontology program that included undergraduate research at ROM. Jess went on to get her PhD at UC Berkeley with the famed vertebrate paleontologist William (Bill) Clemens. Jess considers herself a biologist; she is using fossil mammals to ask biological questions about their evolution and relationships.

Other colleagues, like me, took a less direct path into paleontology, usually starting in college. Arnold Miller (University of Cincinnati) is an academic sibling. Arnie was an undergraduate at Rochester while I was pursuing my master's degree there, but he was very much a member of our group. Another kid from New York City, he liked rocks and minerals and fondly remembers field trips to the American Museum of Natural History. He also had a sixth-grade teacher who was a "geology nut" and took him on field trips. When he entered college, Arnie majored in geology but had no intention of becoming a paleontologist. "Several students on my dormitory corridor talked up geology as an enjoyable major, with courses taught by cool professors." Then he took Dave Raup's introduction to paleontology class, which used the groundbreaking textbook *Principles of Paleontology*, which Raup had written with Steve Stanley. The idea that fossils were not just things but were data that allowed us to answer big questions in the history of life sold Arnie on his career. But he did not go straight to graduate school. Arnie was a journalist for the campus newspaper while in college and considered a career as a science writer. Meanwhile, he worked for a year as a research assistant at the West Indies Laboratory in St. Croix. This convinced him to return to research. Arnie returned to school to do his master's degree at Virginia Tech with Richard Bambach, and then did his PhD at Chicago with Sepkoski. He went on to write, with fellow University of Chicago graduate Michael Foote, a new edition of *Principles of Paleontology*. At Cincinnati, along with David Meyer and Carl Brett, he helped build one of the leading paleontology programs in the country.

A word about Richard (Dick) Bambach. He was part of a remarkable group of Yale graduate students in the 1960s and early 1970s who have influenced paleontology to this day. Dick has published some of the most important and influential papers in the field on issues such as the history of diversity over time and the ecological history of marine organisms. Now in his eighties, he is still publishing today! But what really stands out about Dick is his genuine warmth and infectious good humor. Arnie was incredibly lucky to have him as one of his mentors.

Rebecca Freeman of the University of Kentucky said, "I grew up on a farm, so I was forever picking up rocks and fossils, but somehow never connected that with the idea of being a paleontologist. It wasn't until I was in college that I took the dreaded required science class. I picked geology because I already had a low opinion of all the other sciences. Physical geology made me realize this was an option, and I immediately switched my major and took invertebrate paleontology the next semester." After getting a bachelor's and a master's degree, she started a PhD at Tulane in 1990 but "met with many, many obstacles that many women of the time were probably handling better than I was able to." She ceased working on her PhD and became an untenured instructor at the same institution. She returned to her doctorate studies in 2009, during which time she met Ben Dattilo in the middle of the Utah desert while they were both doing fieldwork. She finally defended her thesis in 2011 and is now an assistant professor. "I am forever grateful to Ben, who really served as an unofficial postdoc supervisor, in that he helped me expand my research horizons and introduced me to so many people in the paleontological community. Ben has been through some rough patches in academia as well, so we sort of bonded over that."

Another later bloomer is Mark D. Uhen, who bounced between geology and biology at the University of Wisconsin at Oshkosh. In the first semester of his junior year, he took paleontology and realized he could do both—he fell in love with the field. Mark did an independent study in college on the evolution of marine mammals. This led him to the University of Michigan to work with the recently retired Philip Gingerich, the world's authority on fossil whales. Mark is his successor for this title. Some paleontologists consider themselves primarily biologists, some primarily geologists, and some a blend of the two; Mark puts himself in the last group. He is now at George Mason University, right outside Washington, D.C., which is one of the few places in the country where you can minor in paleontology. It is also very convenient to the National Museum of Natural History.

Some people's paths are even longer. Growing up in the suburbs of Washington, D.C., Janet Burke admits she was a poor student in high school, even dropping out at one point. Not thinking college was for her, she lived on her own and worked as a receptionist for six years. Hoping to find a place where people cared what you thought, Jana enrolled in night school at the local community college. Her success there led to admission to Smith as an Ada Comstock Scholar, which was designed for "women of nontraditional college age." She ended up taking a class in paleontology with Sara Pruss, who became her mentor. This led to an internship at the NMNH working with Kay Behrensmeyer. Jana became a graduate student at Yale studying the biology of planktonic foraminifera, a group of single-celled organisms critical to our understanding of ancient climates, and she is now a postdoc at Michigan State University.

Although my colleagues took many routes to a career in paleontology, one commonality that stands out is that someone—a parent, a teacher, or a professor—sparked and encouraged their interest. These individuals were willing, indeed eager, to pass their passion on to the next generation of paleontologists and to mentor them. That is the role that Eldredge played for me. Some of these mentors have motivated a disproportionate number of students. A standout example is David Bottjer at the University of Southern California; nearly twenty of his doctoral students can be found as faculty members across the country. One of these students is Mary Droser, now at the University of California at Riverside, who in turn has mentored yet another generation of paleontologists. Lidya Tarhan, currently an assistant professor at Yale, was one of Mary's students and thus a Bottjer "grand-student."

Another name that crops up repeatedly as a mentor is Bill Clemens at the University of California, Berkeley. Clemens, who passed away in November 2020, was the major professor for many important vertebrate paleontologists.[2] Others whose names come to mind are Karl Flessa (University of Arizona), Leo Laporte (University of California, Santa Cruz), and David Jablonski and Susan Kidwell at the University of Chicago.

The Smithsonian National Museum of Natural History (NMNH) has long been a place where young paleontologists get their training and mentoring. Generations of graduate students and postdocs have worked there in the Department of Paleobiology with their large curatorial staff. Their collections are virtually unsurpassed, as is the intellectual vitality.

Some departments are or have been breeding grounds for generations of paleontologists; they produce cohorts of leading scientists that graduate around the same time. One such group came out of Yale in the late 1960s. Another cohort that followed me at the University of Chicago, influenced by Sepkoski and Raup, became informally known as the "Chicago mafia." I suspect that another cohort is the group of doctoral and postdoctoral students currently and recently at Yale. Critical to this program's success is the 2003 hiring of Derek Briggs, who had long been at Bristol University in England. Derek encourages independence among his students, but he has also built strong communal interactions among the paleontology students and faculty at Yale, such as having them over for social gatherings at his house a few times a semester. Derek's close and warm interactions with his students was recognized when received the 2017 Graduate Mentor Award in Natural Sciences from the Yale Graduate School.

So how does one go about becoming a paleontologist? First, an education in paleontology is not a hard, rigid curriculum with a predefined path and set of courses. Some of us are trained in biology and know a lot of geology; others are geologists who are comfortable with biology. We all learn at least some of both. Some of us have received detailed training in the rigors of fieldwork; others have learned cutting edge methods in modeling and data analysis; and still others have learned how to use sophisticated analytical equipment. Most of know at least something about all of these. Paleontologists are by nature multidisciplinary and multitalented scientists; it is one of the things that make it so much fun.

With few exceptions, you cannot major in paleontology. Almost no colleges offer such a major.[3] The same is true for graduate schools; none of my degrees say "paleontology" on the parchment. Budding paleontologists need to major in geology, biology, or best of all, some combination of the two and take whatever paleontology courses are offered. Courses in statistics, geographic information systems (GIS), and programming are very useful. Seize whatever opportunities are available for independent research, either at your university or at a nearby museum, if possible. Don't be shy about asking your college's paleontologist if you can work on an ongoing project. A published paper with your name on it looks great when applying to graduate school.

Outside of class, read as many paleontology books and articles as you can. If your paleontology professor has graduate students, get to know them and see if there is a journal club where you can become familiar with the primary scientific literature. One way to identify a future graduate advisor is by reading a paper you find interesting and talking to the author. You should also join a professional paleontology organization. They all offer heavily discounted student memberships, as well as awards and grants aimed at students. If you can afford it and are capable, take a field-oriented course during the summer—some geology departments require it for the degree, and it is a great experience even if it is not required.[4]

As you approach college graduation, you have to make some critical decisions about whether to go on to graduate school or look for work with a bachelor's degree. Make your decision based on knowledge! I served for many years as my department's director of undergraduate studies, and I was surprised how little college students knew about graduate school. For example, they were not aware that most graduate programs offer both a tuition waiver and some sort of stipend to incoming students. The demands on a graduate student are also different from those on an undergraduate. There are usually fewer formal classes but a high expectation that students will spend time on research. There are also the demands of being a teaching assistant, if that is how you are supported. Late nights and weekends are par for the course.

Going to graduate school can be a long haul, two or three years to a master's degree and at least four years more than that for a PhD. It's long hours, minimal pay, and no guarantee of a job at the end (see the next chapter). Not everyone makes it. There are a lot of positive things, but you still have to *want* to get that advanced degree. Again, if possible, talk with graduate students about their lives. Some jobs in paleontology, such as mitigation paleontology or preparators, don't require an advanced degree, and you have obtained a skill set through your studies that can be applied more broadly in employment.

If you decide to go to graduate school, you are faced with additional choices. The first is whether to go for a master's degree or go directly for a PhD. This depends on a variety of factors. First, a master's degree can be completed in a

few years and does widen the range of possible jobs for which you are qualified. Second, a master's degree provides a great opportunity to become familiar with research and with the science before you face the high demands of a
doctorate program. If your undergraduate program does not include a strong
independent research component, you are probably not ready to go directly
into a PhD program. If do you go on to get a PhD elsewhere after receiving
your master's, you have an additional set of people who can help you with
your career.

Where should you go for a graduate degree? Again, this is where the faculty members in your department are critical resources. They should know
the people in the field and which programs have good (or bad) reputations.
If possible, look at programs with multiple people in the field, so you can be
exposed to a variety of ideas. Do not rely on name recognition for the school!
You will be working with people, not the name (and many programs have
reputations based on faculty who are long gone). There are excellent paleontology programs across the country, both at private schools and at state
universities. Visit the school and have informal discussions with the graduate
students who are already there. Many universities have programs for groups
of prospective students.

Do your homework before you get there, and be prepared with these questions: How much is the stipend? How many hours a week will I be expected
to spend as a teaching assistant? Is there support to attend meetings? How
about research support? What is the cost of living? These questions are not
trivial. There are real differences among programs in how much they pay
graduate students and what benefits students can receive. I lost more than
a few potential students to other programs because they paid significantly
more; some of my department's graduate students take outside jobs because
of the high cost of living in Chicago. This cuts into their ability to spend
needed time on research. There are many advantages to being in a well-
funded program.

At some point, a graduate student needs to choose an advisor. Selecting
an advisor is the most critical decision a budding scientist can make. Your
successes and failures are theirs, and they are your champion when things
get rough. And, like parents, there are good ones and bad ones; I know of not
a few students who left graduate school because their advisors were impossible human beings. As Gould says in *Wonderful Life*: "Many students don't

understand the system. They apply to a school because it has a general reputation or resides in a city they like. Wrong, dead wrong. You apply to work with a particular person." Talk to the students!

Sheila Widnall, then president of the American Association for the Advancement of Science (AAAS), wrote in 1988:

> It is at the graduate level that the student begins to function as an independent scientist—indeed, that is the purpose of graduate education. During this process the faculty gradually begins to remove the props supporting the student and to place more responsibility on the student for problem formulation, evaluation, execution, and defense. . . . These experiences include opportunities to present and defend research results in regular and productive group meetings, to evaluate and criticize the work of peers, to formulate and carry out research tasks of increasing importance, to participate in dialogues and debates about scientific and technical issues, and to discuss future career plans as they relate to current interests and activities. Faculty members often do not make these parts of the educational process explicit to the student. Much of the stress of graduate education results from a lack of student understanding of this hidden agenda.[5]

Good mentoring makes this hidden agenda explicit.

I will add to Widnall's statement that graduate students also need to prepare to succeed in the classroom. The days when professors, even at a research university, are assumed to be able to teach a subject because they are familiar with it are past. Pedagogical methods and teaching technologies are rapidly evolving. New professors (and us old ones) need to be able to effectively teach their subjects. More and more universities are offering programs for graduate students to improve their classroom teaching. Be sure to participate in one; not only will it make you a better teacher, but you will also be much more attractive to potential employers.

Do not neglect opportunities to participate in outreach and informal education, starting when you are an undergraduate. I have been a National Science Foundation reviewer for graduate fellowships and postdoctoral support. Although the scientific merit is essential, the societal benefits (that is, "broader impacts") are also key. You do not have to reinvent the wheel on this; many existing organizations would be pleased to have you work with them.

As a final note, graduate school can be quite competitive. There is an implicit competition for the respect and support of the faculty and an explicit competition for the limited number of positions available. This competition is usually not vicious; the friendship and encouragement of fellow students is critical to success. These friendships will last a lifetime. Becoming a paleontologist is not a lonely path; it is in school that young paleontologists not only learn the science but learn how to become members of a global community.

11

LIVING IN THE REAL WORLD

I n one of my favorite *New Yorker* cartoons, a dispirited looking gentleman at a cocktail party remarks, "My mistake was
going into cosmology for the money." Of course, none of us
went into paleontology for the money (if you did, you were sadly
misled), but we still need to make a living. One of my retired
colleagues would become livid if someone suggested that professors did not "live in the real world," with the implicit assumption that those of us in academia were somehow insulated from
the struggles of "real people." Nothing could be further from the
truth. Although our topics of study may seem esoteric, scientists
still must pay for the necessities of living. We shell out money
for rent, buy houses and cars, marry and divorce, send our children to college, and worry about health care and retirement. The
implicit assumption is that after years of college and graduate
school, there should be at least a modicum of financial security
waiting for us. The path to that is long, laborious, and full of
dangers and disappointments, with many falling by the wayside.
When the goal is reached, however, the reward is one of the best
jobs in the world.

The next step for many after earning a PhD is the postdoctoral
fellowship, or postdoc. Postdocs enable you to continue your

training in research while having time to publish all the papers and develop the grant proposals that will allow you to land a permanent position. Postdocs generally pay on the order of $40,000 to $50,000 a year. You are not expected to teach, but postdocs usually put in long hours on research and writing, well above what would be considered overtime in many fields.

Postdocs are highly competitive, and there are simply not that many of them. A key funding source in the United States is the National Science Foundation, which funds a successful applicant for two years, either as an Earth Sciences Postdoctoral Fellow or a Postdoctoral Fellow in Biology. As of July 2019, about four active fellowships fell in the broad area of paleontology. Another government agency with a smattering of postdocs is NASA, which sometimes funds researchers on the origins of life through its exobiology program. A small number of institutions, such as the Smithsonian Institution and the Florida Museum of Natural History, have endowed internal funds to support postdoctoral scholars. Some doctoral students can find postdocs in other countries, notably in Germany and the United Kingdom. Some highly qualified young paleontologists receive successive postdocs, but this means uprooting yourself and perhaps a spouse or family every two or three years.

First cousin to the postdoc is a soft money position. "Soft money" means you are supported by a grant, often working under a more senior investigator. This contrasts with hard money, which means you are part of an institution's regular budget, such as is the case with a regular faculty member. If you are on soft money, your primary research is constrained by the terms of the grant award. Leanne Elder was on such a soft money position at Yale under various grants to Pincelli Hull. A biologist who studied the physiology of marine invertebrates for her PhD, Leanne originally worked as Hull's lab manager, which was transformed into a postdoc position. She has extended her studies now to the physiology of foraminifera, a key group of single-celled organisms in the modern ocean, and uses this work to improve interpretations of the fossil record. In 2019, Leanne moved on to become collections manager for the Division of Invertebrate Zoology at the Colorado University Museum of Natural History.

Another somewhat less appealing option is a temporary teaching position. Many institutions hire faculty on an as-needed basis to cover classes that need to be taught. This may happen if a regular faculty member retires, leaves for another institution, or is on sabbatical. What they don't offer is tenure, the near-guarantee of lifelong job security. At best, these full-time positions pay a living wage, are relatively secure, and you are treated as a regular faculty member. It is just as likely, however, that you must scrape by financially, be constantly worried about keeping your job, and barely be acknowledged. At worst, you could become an "academic nomad," forced to teach part-time at multiple institutions for a pittance. These types of positions are becoming increasingly common because a temporary hire or a nomad is much cheaper than a permanent faculty member. The American Association of University Professors estimated that more than 70 percent of all faculty in American higher education are non-tenure-track "contingent faculty." As expected, most contingent faculty have little or no loyalty to the institution or its students and cannot be on necessary committees or share in governance.

For most, but not all, young scientists, the dream is to land a tenure-track position at a college or university. These institutions range from major research universities, to small liberal arts colleges, to local community colleges. Each of these institutions is looking for different sets of abilities in a job candidate. At the most prestigious and competitive institutions, someone with a great track record of publications and a documented ability to get outside funding, such as a grant from the National Science Foundation, is desired. Teaching ability is less emphasized. In contrast, a smaller institution, such as a private liberal arts college or a regional state university, might want some demonstration of research competence but will be more interested in teaching proficiency. My former PhD student Karen Koy landed such a job at Missouri Western State University, where she is tenured and has won awards for her teaching. Another former student, Chris Ebey, is happily ensconced with tenure at a junior college in upstate New York. She also recently received a teaching award. Some of these positions do come with punishing teaching loads that all but totally preclude research, making them quite unattractive to some.

For those who are ambitious to make a mark in the field, a four-year institution that encourages or even requires top-notch research is the objective. At most half a dozen positions open every year, and competition is fierce. This small number of positions is partly due to the erosion in the number of institutions that have even one paleontologist on their faculty. Some of the advertised positions are not specifically for paleontology but are broad enough that a paleontologist can fill them.

Jacalyn Wittmer is an invertebrate paleontologist with a PhD from the Virginia Polytechnic Institute. She describes the search for employment as a "mental rollercoaster of disappointment, self-doubt, patience, and persistence." She did not look for a postdoc because she did not want to uproot herself and her family every two years. She also recognized that getting a job strictly as a paleontologist limited her opportunities, so she also marketed her skills in stratigraphy. Jackie eventually found a position as a lecturer, a non-tenure-track teaching position, at the University of Illinois at Urbana-Champaign (UIUC), temporarily replacing a retired senior and distinguished paleontologist. Unfortunately, she found that UIUC "was not willing to have tenure-track positions in 'classic' paleontology anymore. To some departments, it seems old and archaic." After teaching a wide variety of courses for three years as well as pursuing her research, Jackie finally received a tenure-track position in paleontology at the State University of New York at Geneseo.

Sometimes it's difficult to even get your feet on the first step of the ladder. I know that some bright young paleontologists gave up the struggle, even as far back as graduate school. Jim Lehane, who has a PhD from the University of Utah, is hunting for an academic job in paleontology. In his case, he did not apply for a job while working on his dissertation because he was concerned about "having my attention divided by having a new job while still trying to finish up my doctorate." During the last year of his doctorate, he was no longer funded by his institution and had to take a job as a specialist in geographic information systems. Jim also did not pursue a postdoc, for reasons similar to Jackie Wittmer's: "My wife is a pediatrician, and it takes time for her to build up a practice. She cannot afford to move to a place for two years, only to move on again afterward." Jim's PhD advisor had also recommended against it. Since then Jim has applied for faculty jobs with limited success. He has found that "out of everything an academic job applicant has to go through in order to get a job in academia, the one most frustrating thing is the complete lack of caring the universities seem to have toward their applicants. Here the applicant spends hours and days perfecting an application, and often the university will not even acknowledge the application, nor inform you when, no, you are not their choice. The best thing one can do is 'suck it up' and move on. But each and every time the rejections hurt, and the lack of acknowledgment hurts even more." (I find this indifference unprofessional and rude.) Having a child with special needs, Jim has now decided to be a stay-at-home parent. Meanwhile, he is working on improving his application and on honing his skill set.

Ellen Currano, a paleobotanist whose PhD included fieldwork in the Bighorn Basin of Wyoming, was much more fortunate. After a year as a National Science Foundation (NSF) postdoc doing fieldwork in Ethiopia, she became an assistant professor at Miami University. "Three years into my time there, the University of Wyoming advertised for a paleobotanist who works on deep time macrofossils, preferably from Wyoming. The job ad seemed written for me—I flew from Ethiopia to Laramie to interview, and thankfully I had an offer waiting for me when I touched ground again back in Africa."

An issue that has emerged in recent decades is that of a two-career couple: the difficult "two-body" problem. The two-body problem is not an exercise in basic physics; it is the difficult challenge of finding employment situations where both members of the couple have meaningful employment at the same place (or at least in the same vicinity). Lucy Edwards (United States Geological Survey) stated that "it's still difficult for a couple to find jobs for both in a reasonable commuting area. If one has a better opportunity at a different location, it's hard for the other to accommodate." The solutions vary greatly among couples, depending on their respective stature and appeal to possible employers, their individual career goals, the nature of the job market, and the culture of their institutions. The resolution of this problem is often revisited and is frequently the subject of deeply personal negotiations as both members of the couple follow their career tracks.

Lidya Tarhan (Yale University) agrees that it can be challenging to find positions for both halves of an academic couple. She feels extremely fortunate that she and her husband have each found faculty positions at the same university and in a department that allows each of them to pursue their scientific interests and careers to their full potential.

Periods of living apart are typical. Susan Kidwell and David Jablonski met while both were graduate students at Yale. A few years older, Dave got his PhD in 1979 and went on to postdocs at the University of California, Santa Barbara and UC Berkeley, finally ending up at the University of Arizona in 1982 where Susan had been on the faculty for a year already. During this period, they had lived apart for two years, and Susan considers it "dumb luck" that they found positions at the same place. A few years later, they both were hired by the University of Chicago, where they have had illustrious careers.

Many times one member of the couple will be the "lead" and the other "the follower"; that is, one member of the couple is offered a position and some sort of accommodation is found for the spouse. (Currently, many institutions will assist a married couple but not those living together; many others balk at a "spousal hire.") Pincelli Hull found herself in that situation when her husband was hired at Yale before she had finished her PhD. She secured a postdoctoral position there, and her superb work led to her being offered a tenure-track position.

Pedro and Katherine Marenco met when they were both graduate students at the University of Southern California, with Pedro being two years ahead. They decided that whoever got a job first would be the lead. That ended up being Pedro, who was hired at Bryn Mawr College. When Katherine arrived there, she had an NSF postdoc. Over the next period of years, she was hired to teach paleontology as an adjunct but occasionally went without pay. This has been very rough on her, especially in terms of what she legitimately expected and wanted in her career. Katherine has finally been hired, but as a non-tenure-track lecturer.

S. Kathleen Lyons and Peter Wagner have taken turns being the lead and the follower. They met about the time he was finishing his PhD at the University of Chicago. Peter then departed for a one-year postdoc at the Smithsonian, returning to Chicago to take a curatorial position at the Field Museum in 1997. Kate finished her PhD in 2001, then went off to consecutive postdocs at the University of New Mexico and the National Center for Ecological Analysis and Synthesis in Santa Fe. Overall, they were apart for more than five years. She and Peter finally rejoined at the Smithsonian in 2007, where he had received a job as a curator. Kate, on the other hand, was supported on soft money, during which she produced a string of high-impact papers. This led to her being offered a tenure-track position at the University of Nebraska, with Peter receiving a tenure-track position as a spousal hire. Peter says the decision to leave the Smithsonian was easy because the museum declined to counter the offer.

Dena Smith (National Science Foundation) views the two-body problem as particularly difficult if the lead is a woman and a person of color. Because the woman is the "diversity hire," it is perceived that there is less need to hire the spouse. There was also implicit questioning of her husband's abilities because he was not the lead. In their case, they made the joint decision to leave the University of Colorado and join the NSF.

So there you are with the job you always wanted: assistant professor at Euphoria State. This is no time to relax, you must get tenure! A word about tenure. Probably nothing in academia is more misunderstood nor more controversial outside of academia. The major upside of tenure is that once it is awarded, usually at the end of the sixth year of employment, it is very difficult to be dismissed except in cases of incompetence or unprofessional behavior. The National Education Association estimates that about 2 percent of tenured faculty lose their jobs for those reasons every year.[1] Another key advantage is that you cannot be dismissed for having controversial ideas. At a time when core paleontological concepts such as climate change and evolution are under attack, having tenure is comforting. Receiving tenure is also not an excuse to kick back and relax. Most tenured professors, myself included, work as hard after tenure as before. Although some people abuse the system, there are ways other than outright dismissal to make their lives unpleasant.

Most important, tenure is an uphill battle. Although the exact criteria differ among institutions, the one common standard is "excellence" in some balance of "teaching, research, and service." In other words, you must excel at the activities that the college or university values. For teaching-oriented institutions, you must demonstrate outstanding and demonstrable abilities in the classroom, perhaps with some public service and research activity. In a research university like mine, research accomplishments must be first rate. This roughly translates to a *minimum* of about two papers a year, an NSF grant, and stellar letters of support from senior colleagues at other institutions. Teaching should at least be adequate; stellar teaching will help but is not sufficient. Service, which refers to time spent on department or university committees, should be minimal. It is a time suck that really doesn't help in the end. Let the folks with tenure do it.

The amount of time spent working in order to get tenure can be killing: eighty, ninety, or even more than one hundred hours a week are not atypical. The best description I have read of the pressures on an untenured faculty member is Hope Jahren's best seller, *Lab Girl*. She had to spend nights and weekends at the laboratory to produce the research to write the papers that would justify the grants that support the laboratory. Repeat as often as necessary until tenure is received and often long afterward. Observing these pressures can lead young scientists to leave the field; Jahren mentions one young student who walked out saying she could not imagine living that kind of life.

The other killer is the uncertainty. For six years, you can be dismissed for any reason. And if you are denied tenure, which happens in about 20 percent of the cases, you are out of work, often after a grace year. I speak from painful personal knowledge here. When I came up for tenure, I had enough papers, well-received teaching, and service to the profession and department. What I did not have was an NSF grant, only smaller grants from private foundations. When the various campus committees looked over my case, they voted me down, albeit by split votes. Eventually my case ended up on the desk of the chancellor. It was at that time that my colleague Karen Prestegaard and I heard we were successful with an NSF grant proposal. I immediately contacted the chancellor, and as the car dealers in New York say, "money talks, nobody walks." He then ordered my case to be reconsidered. I thought this meant an immediate review, but the powers that be decided I would have to go through the entire process from scratch. I went through unanimously this time, but it was more than a year before it was official. Meanwhile I was on a terminal contract and had a young child. Worst year of my life. Decades later, it still rankles.

I am not the only one to experience this. To the shock of many of us, a colleague was denied tenure at a prominent liberal arts college. This was a paleontologist with an active and respected research record. Thankfully, he landed an excellent position at another institution.

Despite all the pitfalls on the way, there are few jobs as rewarding as being a tenured faculty member. Of course, the job security is a large share of that. Unless you misbehave, or the college is financially failing, you have a job for as long as you are willing to work.[2] To me, however, the best part is the freedom to pursue whatever research direction I find interesting. My grandmother once asked me, "Who tells you what to work on?" And the wonderful answer is, "No one!" I choose my research topics, sometimes several at a time, and pursue them without worrying that someone will stop me. Of course, this freedom has some constraints. I am expected to publish my research in reputable journals at about the same rate I did before tenure. I am also expected to pursue relevant funding opportunities to support my work and that of the institution (I return to this later). If my research were of inferior quality, I would not be able to publish or get grants.

Research is one leg of the triad of "research, teaching, and service." Everyone in my department teaches, even us white-haired full professors. Some of this teaching takes place in the classic classroom setting, where I and my colleagues can instruct everyone from first-year students in large lecture classes to advanced graduate students in intimate seminar classes. Other teaching takes place outside the classroom, mentoring undergraduates, advising graduate students, and serving on graduate student committees. Some of this instruction can be quite informal; I recently taught a graduate student how to snorkel in preparation for fieldwork at a marine lab. The amount of formal classroom instruction required, the so-called teaching load, varies greatly among institutions. At a research university, one or at most two courses per semester is fairly typical. At small liberal arts colleges, six or even eight courses per year are not unusual. Faculty at these institutions have to be creative in finding time to pursue research. This is unfortunate; as my late colleague Robert Demar pointed out, teaching and research are synergistic. I find that teaching a course and updating it forces me to keep current in my discipline. At the same time, being an active researcher means that I can teach the state of the art in my field. And it sometimes leads me to develop new courses that cover these hot topics.

The third leg of the triad is service, which has until recently been the least valued, especially for promotion and tenure. This can include service to the department or university, such as serving on committees; service to the profession, which includes reviewing papers and grants and serving as an officer in a professional society; and service to the community, which covers a wide range of activities including public presentations or classroom visits. These activities are best delayed as much as possible until after tenure. Although untenured faculty should do some service, it is usually only a minor contributor to the tenure decision. I take service seriously, especially to my professional society. I view it as giving back for the opportunities created for me by previous generations of scientists.

The combination of these activities means that my days, and those of my colleagues, are full. When I talk with my wife over dinner about what we did today, it is rare that I discuss the same thing. Every day is different. It is also why attempting to make an academic fill out a time management chart is doomed to failure. We are judged by what we accomplish, not by how much time we take to do it.

A positive shift in recent years has been the recognition that young faculty scientists need to be mentored. It is not enough for them to know field and laboratory techniques; they also need to learn the skills, including teaching skills, that will help them successfully navigate their careers. Fortunately, universities and scientific organizations have recognized the need to provide young scientists with the tools they need to succeed. At my university, every young member is assigned a mentor, and there are workshops on grant writing and teaching methods. Most scientific societies have programs that provide support to young investigators, including opportunities to meet and be mentored by established scientists.

One of the best programs is administered by the Science Education Resource Center (SERC) at Carleton College. I took a workshop from them on course development that totally changed how I teach. Their website is full of teaching strategies and concepts. Most important, they run a multiday Workshop for Early Career Geoscience Faculty, which covers "effective teaching strategies, course design, establishing a research program in a new setting, working with research students, balancing professional and personal responsibilities, and time management." I wish it had existed when I was starting out; I could have avoided many mistakes.

Looking forward, probably the most challenging aspect of the job search in academic paleontology is the erosion in the number of positions. When I started out at UIC, I was one of four paleontologists; there was a vertebrate paleontologist in my department and a vertebrate paleontologist and a paleobotanist in biology. As faculty retired, they were not replaced. Similarly, the invertebrate paleontologist at our campus in Urbana-Champaign has not and may never be replaced. As Jackie Wittmer says, part of this is the perception among our colleagues in geoscience that paleontology is "old science," that is, no longer cutting edge. Related to this is that most paleontological research is cheap. As Jacques Gauthier told me, most of us can do our investigations for about $10,000 per year, so grants to paleontologists bring in a fraction of the overhead (fraction of the grant that goes directly to the institution) of other disciplines, such as geochemistry. This makes hiring a paleontologist much less attractive to deans and department heads. Many paleontology slots have turned into "geobiology." Biology departments have seen similar shifts as natural history biologists (organismal biologists) are replaced by specialists in the more lucrative areas of cell and molecular biology.

In a 1997 overview of academic paleontology in the United States, Karl Flessa and Dena Smith noted that most institutions that had a paleontologist had only one of them ("the typical lone paleontologist").[3] I did a follow-up study in 2007 and identified about one thousand paleontologists of all types in American academic institutions.[4] A couple of results stood out. First, the field seemed to be aging. There were relatively few assistant (pretenure) professors. Many of the recently hired faculty work on Neogene climates and environments and are attractive hires to departments refocusing on global change and related environmental science issues. How about the "lone paleontologist" pointed out by Flessa and Smith? As of 2007, roughly half of the 350 four-year institutions that had at least one paleontologist had only one paleontologist. For universities alone, the percentage is 44 percent, up from the 34 percent figure found by Flessa and Smith. It is only in museums and a few university departments that large concentrations of paleontologists remain. I suspect that a more recent survey would show that these trends have continued.

Finally, I would be remiss if I left you with the impression that all paleontologists work at or want jobs at universities. As I have discussed elsewhere in this book, multiple other career paths exist: at museums, in government, and in the private sector. Each of these career paths has its own opportunities and challenges. Beyond this, as Arnie Miller suggests, the future of the profession "transcends positions in academia." Not only are paleontologists well equipped for jobs away from college and universities, but they also have valuable skill sets that can be readily applied in other disciplines. Bruce Mac-Fadden at the University of Florida teaches a class called "Alternative Careers in the Natural Sciences." My former PhD student Tao Zhang did his thesis on the applications of artificial intelligence in paleontology. He has since mapped out a career in the tech industry. The students of Michal Kowalewski (Florida Museum of Natural History) use their paleontology skills doing coastal environmental assessment. Paleontologists know a lot about a lot of things and can do a lot with what they know!

12

THE FACE OF PALEONTOLOGY

At the 2014 North American Paleontological Convention (NAPC) in Gainesville, I made plans to go to dinner with my good friends and colleagues Lisa Park Boush and Dena Smith. As is inevitably the case at such times, others soon joined our group. When we finally sat down for dinner, I was the sole male, with seven female colleagues. I was flabbergasted.

To understand why I found this so dumbfounding, I need to mention that in my four years of college and six years of graduate school in the late 1970s and early 1980s none of my professors were women. This includes not only the paleontologists but all of the geology faculty. There was a scattering of female students in my classes, and during my first two years of graduate school, there was only one female student among us. In my early days of attending conferences, I recall very few senior women scientists, although there were some outstanding ones such as Jennifer Kitchell, Elisabeth Vrba, and Anna Kay Behrensmeyer. If you went to dinner, the odds are that there would be a crowd of men and one or two women. The dinner at Gainesville brought home something I and other paleontologists of my generation intuitively knew; there has been a major demographic shift in paleontology over the last several decades. The NAPC itself shows this change.

The NAPC is a gathering of all flavors of paleontologist, and it has been held only eleven times since 1969. At that 1969 meeting in Chicago, there were eighty invited talks; only two were given by female paleontologists. There were special "Ladies Events" that included shopping and a tea (the men had a "smoker"). In contrast, at the tenth NAPC in 2014, 40 percent of the speakers and 60 percent of the authors were women.[1] Female participation in the science has clearly increased over the last fifty years.

Although there have always been female paleontologists, for most of the history of the field they were a distinct minority. One heroine is the Englishwoman Mary Anning (1799–1847), who spent many years collecting fossils among the spectacular Jurassic cliffs along the shore in Lyme Regis. She amassed and prepared many important fossils, which were sold in her small shop and provided a livelihood for herself and her family.[2] Among her most important finds were complete ichthyosaurs, pterosaurs, and the first plesiosaur.

Unfortunately, despite the ground broken by Anning, the number of professional female American paleontologists active at any time over the rest of the nineteenth and most of the twentieth centuries could be counted on the fingers of one hand.[3] On display at the headquarters of the Geological Society of America is a poster with all its members in 1899. Amid all the bearded worthies is Florence Bascom, professor of geology at Bryn Mawr College and the sole woman. Although not a paleontologist, Bascom mentored Julia A. Gardner, who went on to a distinguished career at the United States Geological Survey, working mostly on Cenozoic mollusks. Gardner in turn mentored Winifred Goldring, who became the first female New York paleontologist and was the first woman elected president of the Paleontological Society in 1949. The Association for Women Geoscientists and the Paleontological Society cosponsor a scholarship in her honor for the outstanding female student pursuing a career in paleontology. Goldring was followed in the post of Paleontological Society president by Gardner in 1952. No other woman held that position until 1985 (Helen Tappan-Loeblich), and only two others have done so since then.

Mignon Talbot received her PhD from Yale in 1904 and went on to become professor and chair of the Geology Department at Mt. Holyoke. She discovered a small Jurassic dinosaur in the vicinity of the college. In 1920 the Yale paleontologist Charles Schuchert wrote his opinion of Talbot and the two other women who had earned geology doctorates at Yale by that time:

As students, women compare well with our men candidates for the higher degrees. As geology is primarily a field science, naturally women do not easily work into it from the standpoint of research in the field. On the other hand, as teachers in women's colleges and in normal and high schools, and as laboratory workers, draftsmen, and assistants in museums, many women will find opportunities for geological work. In the Government survey and in some of the state surveys they are employed as paleontologists, statisticians, librarians, etc. As yet, women have not in numbers gone into geology and paleontology, but they are more and more entering into the study of these sciences in the universities, and, so far as one can see, although the field is a limited one, they are all finding places when they choose to adapt themselves to the work demanded of them.[4]

This quote reminds me of two mid-twentieth century chemistry sets that I have in my office. One is the "Gilbert Chemistry Outfit for Boys," with a display of the boy and the scientist he will become. The other is the "Gilbert Lab Technician Set for Girls," and yes, it is pink.

These attitudes were still openly expressed in the 1960s and 1970s, but early signs of change were in the air. Kay Behrensmeyer (PhD, Harvard, 1973) recalls that "I came from a family with strong women who gave me a pretty unshakable sense of self-worth from the beginning, and this helped me ignore attempted put-downs and gender bias in college and graduate school. A low point was when a senior professor let it be known that the only reason women came to Harvard for a graduate degree was to 'catch a Harvard man.' I simply ignored him—the statement was so ludicrous." She adds, "On the bright side, I had male teachers and peers from undergraduate school onward who supported me and treated me as an equal—they were on the forefront of championing a more even playing field for young women entering geology and paleontology as a career choice." Even today, however, "the main problem has always been being taken seriously, or even just being heard, especially by male peers or more senior people. Since I'm a senior myself these days, the latter is no longer an issue, but sometimes I realize that administrators or others not familiar with my work still require extra assertiveness on my part. Occasionally I'm surprised at a comment or other hints of dismissiveness on the part of male colleagues. But fortunately, it's pretty rare these days."

Patricia Kelley, who went to graduate school at Harvard after Behrens-meyer, recalls that "there was definitely a hostile air—the 'bullpen' that held desks for several male students of Kummel and Gould had pornographic posters of women covering the walls. I was given my own office rather than being placed in the bullpen."

Ellen Thomas is a micropaleontologist who earned her PhD in 1979 at the University of Utrecht, Netherlands. She was the first woman to get a geology-based PhD at Utrecht. Ellen has also seen significant changes in the status of women in the field since she was a student in the late 1960s: "As a student, I was told I was not welcome and should do something else than studying geology—[I was] often told that women should not take jobs from men who need them to support a family." Her professors were not supportive. There had been a number of female micropaleontologists in a generation before hers in the United States, but none in the Netherlands.

It is difficult to get precise numbers on the growth of women in the pro-fession. A study estimated that in 1977 the Geological Society of America had between six hundred and one thousand women among its twelve thousand members. The proportion for the Paleontological Society at that time can rea-sonably be estimated to be the same; about 5 percent of the papers at the 1977 NAPC had at least one female author. In the past forty years that proportion has grown; a current estimate by Phoebe Cohen, Alycia Stigall, and Chad Topaz is that women (those with female-gendered names) make up about 40 percent of the membership.[5] Interestingly, there is a strong age bias in that overall number. Women make up 50 percent of the students but only 22 per-cent of the professional members. Numbers are apparently similar for the Society for Vertebrate Paleontology. Part of this disparity may be due to the influx of younger women into the field, but some, as Alycia Stigall suggests, may also be due to students leaving the field before they become profession-als. This attrition has been well-documented across the sciences.

Much has been written about why women leave or decide not to pursue careers in science. I can only briefly and inadequately touch on it here. One oft-discussed issue is the competing demands of a research and teaching career versus family life. Decades ago, most scientists fit classic gender expec-tations. The researcher was invariably male, and his spouse was a housewife. She would be expected to take care of all the domestic needs, including child rearing, while her husband spent long hours, often well into the evening or weekend, in the laboratory. In the case of geologists and paleontologists, this might instead mean long stretches away from home in the field. Either could put a strain on marriages. The concept of life-work balance was unknown. Divorce was rife among my professors. My advisor died from a heart attack at age fifty; another committee member dropped dead at the age of forty-eight. Not good role models.

With the advent of more women in science and of two-career couples in general, there have been major alterations in these patterns. It is often the case that a woman is the scientist and the spouse is not, or that both members of the couple are scientists, or at least academics. Expectations regarding who is responsible for home life and child care are also shifting. Increasing attention is being paid to life-work balance, with many younger scientists questioning the number of work hours implicitly expected.

Decisions about if or when to start a family have become much more difficult. Universities have not yet adjusted to the conflict between the tenure clock and the biological clock. In 2002 Carol de Wet, Gail Ashley, and Daniel Kegel pointed out that the schedule for promotion and tenure was set up decades ago by and for men.[6] Even in ideal circumstances, tenure is not usually granted to individuals younger than their mid-thirties. As a result, "there is an unavoidable collision between a women's optimum child caring years and her trajectory." They suggested that this was a major cause of the attrition of women from the geosciences prior to receiving tenure. In a similar vein, Wendy M. Williams and Daniel Ceci wrote in 2012 that "for women, the tenure track presents a harsh reality that juxtaposes the most significant physical and emotional challenges of their lives with the most significant professional challenges."[7] They identified the desire for children and family life as the most important factor in why women were underrepresented. In a 2017 article in the *Atlantic*, Nicholas H. Wolfinger drew on many of the same studies cited by Williams and Ceci to argue that "family formation—marriage and children—represent the biggest stumbling block women face *en route* to a career in academic science."[8] He again points out that it is the current rigidity of the academic structure that "doesn't offer women any good time to have children."

Many women paleontologists have had to rely on supportive spouses and other family members. Patricia Kelley, who recently retired from the University of North Carolina Wilmington, says that "it is possible to have a successful career in the geosciences without giving up the joys of marriage and children. (It's not easy but it can be done, especially with the support of family and colleagues.)." In her case, her minister husband committed to equal participation in child rearing and to "following his wife's career." This type of relationship is becoming more typical. Kate Bulinski's (box 12.1) husband is a teacher; for them, with four young children, "the rhythm of life is the rhythm of the school year."

BOX 12.1

Kate Bulinski was one of Arnie Miller's doctoral students at Cincinnati and is now a tenured faculty member at Bellarmine University, where she was the first person to teach geology at that small liberal arts school. A self-described "army brat," Kate was always interested in science but not particularly in paleontology. She started as an undergraduate at Penn State but changed to geology after taking a class in it. It was a field course with Mark Patzkowsky that convinced her to go into paleontology and eventually to graduate school. Kate's success is even more remarkable because of her struggles with health issues while in graduate school. *

---

*W. Beckman, "Kate Bulinski—A Great Find," University of Cincinnati, June 4, 2008, http://www.uc.edu/profiles/profile.asp?id=8518.

The issue of fieldwork or other travel remains a major stumbling block. Susan Butts (Yale Peabody Museum) states that "I travel a lot, but as a woman with a family, that's hard." As Phoebe Cohen (Williams College) puts it: "I would say one of the special challenges of paleontology is that there is often a fieldwork component. This can be tough for families with young children, women who are having kids, and people with less than supportive partners when it comes to ditching responsibilities for weeks or months each year. The age of men heading off to the field for the summer while their wives stay home and take care of the kids, house, etc. is coming to an end—though some people certainly still have that arrangement, and some women do have partners who will take on that role for them." In some cases, a grandparent (of course, usually a grandmother) pitches in. Kay Behrensmeyer concurs: "A continuing challenge for women paleontologists, and scientists in general, is how to balance a career involving fieldwork or long hours in a lab and classroom (or both) with family. Some universities are developing positive strategies for spousal hiring, which is a good sign, but the pace needs to pick up so that more women can stay in science rather than dropping back or dropping out to have a family." Even if there is a willing partner, I will add that leaving your children and partner behind while you go into the field or attend a conference can still be quite stressful. Once their children are old enough, some of my colleagues bring their kids with them when they do fieldwork or go to a meeting or even to the office or lab. The last is necessitated because most institutions have not as yet recognized the need for child care, especially for younger children. Alycia Stigall and Kate Bulinski have stressed the "need to be efficient" in their time management.

Recognition by institutions that systematic changes need to be made to address these issues is slow in coming. When Tricia Kelley had her children in the 1980s, her institution did not offer maternity leave: "I was the first and only woman faculty member in the School of Engineering at the University of Mississippi during my time there, and there wasn't even a women's bathroom for me to use! I tried to avoid telling my colleagues I was pregnant as long as possible because I knew what their reactions would be—they assumed my career would be over. My chair asked me in March if I could postpone having my baby from August to October because it would help the department out in juggling the teaching schedule."

Changes have been made in many places. The 2002 article by de Wet and others called for stopping the tenure clock and allowing part-time work for a given period. Williams and Ceci included those suggestions but also made numerous recommendations, such as working from home, reduced teaching loads, and university-based child care. Wolfinger also supported part-time options. As Wolfinger pointed out, it is important that this is an entitlement rather than an option that must be requested and approved. Otherwise, there might be reluctance to apply for the hold for fear of not looking dedicated enough. Just as important, this hold should apply to both men and women and to same-sex partners.

Having children is not restricted to faculty; graduate students are also in their childbearing years. Wolfinger suggested that family friendly benefits also be made available to graduate students. In 2017 a group of Yale graduate students went on a hunger strike; affordable child care was one of their demands. Anne Weil states: "Family leave is necessary for graduate students. Advisors need to change their attitudes about students having babies." She also notes that we need to be cognizant that in many cultures, such as among Native Americans, family comes before anything else.

Impacts on family life are not the whole story of why young women disappear from science. Many of my female colleagues have not married or have decided not to have children, despite familial or societal pressure to do so. Meg Urry, past president of the American Astronomical Society, is not convinced by the "childbearing years" argument. She stated in 2010 that "it is certainly true that too few high-quality child care options are available, and that women do more family care than men do. But women without children still do not advance at the rate men do."[9] In a note to me, Urry points out that "women have become a large part of some very demanding professions, such as medicine, where trainees have to be on call and where hours are long and, in some cases, quite rigid. Women's participation also varies widely among different disciplines. This suggests it is local culture in a discipline that governs women's participation, not their family plans." She adds: "Academia is a great career for women who want to have families, as it provides a degree of flexibility (and, increasingly, parental leave and stop-the-clock tenure policies) that is not available elsewhere. That doesn't mean it isn't important to improve our workplaces, to make them more compatible with family life. Young men will be grateful for this as well." Urry summarizes the problem of equity as "our perception of women being less good than men, when objective (gender-blind) reviews say otherwise."

Self-image issues are also possible. Imposter syndrome, broadly speaking, is when someone who is a scientist, or is in the process of becoming one, has severe self-doubts about not being competent enough to succeed and is afraid of being found out as an "imposter." Alycia Stigall and Rebecca Freeman were organizing a session at a national GSA meeting when they "encountered extreme difficulty in acquiring women to accept an invited speaker position for a session."

We had to ask five women to get one to finally agree. Conversely, the first two men both immediately agreed. The more interesting part is that the first three women indicated they didn't feel qualified to speak and so were declining. One of the men made a similar comment but accepted anyway. It was an interesting reflection on the known gender bias of women having lower confidence overall and declining opportunities or waiting to seek promotion versus less qualified men actively seeking promotion. It was kind of crazy to watch it play out.

There are also stresses in the classroom. Women professors are, in general, not treated with the same respect as male professors, especially in the sciences and engineering. Kate Bulinski, who teaches at Bellarmine University, mentioned that she has had authority issues with students because she was young woman; male athletes especially tend to get in her face.

The elephant in the room is sexual harassment, especially of women students. I discuss such inappropriate conduct at conferences and during fieldwork in other chapters. At this point, I will just mention that a blind eye was turned to such behavior throughout my student career; it can now deservedly lead to dismissal.

A separate but not unrelated issue is the lack of recognition of the scientific accomplishments of women paleontologists. At its annual business meeting, the Paleontological Society announces a variety of awards. The most prestigious of these is the Paleontological Society Medal, offered for lifetime achievement in the science and awarded since 1963. Only four women have received this honor. In 2018 the great Kay Behrensmeyer finally received the medal; it was long, long overdue.

A similar award, the Romer-Simpson Medal, has been given by the Society of Vertebrate Paleontology since 1987 and has had only five female awardees. The Paleontological Society also gives an early career award, the Schuchert Award. Six women have won since its inception in 1973. As Alycia Stigall (who was awarded the Schuchert in 2016) pointed out in a 2013 essay, "women are not being recognized proportionally to their participation."[10] This can be discouraging to a younger generation of women scientists. Phoebe Cohen notes: "While times have changed, one need only look at the record of PS medal and Schuchert awardees over the last 30 years or so to see that there is still room for progress."[11] In 2020 women won both awards, as well two other significant awards given by the society.

Efforts have recently been made to raise the image of women as paleontologists. PaleoFest, an event held annually since 1999 at the Burpee Museum in Rockford, Illinois, was organized by its then science director, Scott Williams. This public outreach event has focused on dinosaur paleontology since its inception. The theme of the 2016 PaleoFest was Women in Paleontology, which featured twenty-seven presenters in two days, all but one of them female (the one male presenter was from Egypt, where he was training that country's first women paleontologists; one of whom spoke at the 2019 PaleoFest). By all accounts, the presenting scientists were thrilled at the large crowd that attended. A similar but smaller event, Women in Paleontology Day, has been held every year since 2014 at the Orlando Science Center in Florida.

The entertainment industry has sometimes supported the image of women paleontologists. One of most exciting things about the original *Jurassic Park* movie was not the incredible CGI dinosaurs (though they were cool). Nor was it the hero paleontologist Alan Grant. It was the paleobotanist Ellie Satler, played by Laura Dern. What made Satler exciting was that she was not there as eye candy; in her scenes, she uses her scientific knowledge, for example, to discover which plant makes a *Triceratops* sick. This is in sharp contrast to the two female paleontologists Ross dates in *Friends*. Their portrayals are insulting: a beautiful Black woman paleontology professor who dates Nobel Prize winners (and Joey and Ross) and one who was "hot" and a slob. We are never told what they actually do as scientists. They are defined by their relationships, looks, or race (the series had been criticized for being too white) and not by their abilities.

On the other hand, documentaries about paleontology have nearly always featured men, often with beards, as their talking heads. I have a beard, and when I dress for the field, I am usually wearing worn jeans and a large floppy hat. Thus I match the longtime gender stereotype of what a paleontologist should look like.[12] As you may have noticed as you read this book, many paleontologists would not naturally have a beard. One night over dinner, Ellen Currano expressed to her friend Lexi Jamieson Marsh the thought that "if only I had a beard, my life would be so much easier." It had been a rough day for her:

One of those where I felt like my predominantly male colleagues did not value my ideas or take me seriously. And I was further discouraged because paleontology had been recently featured in several major popular science venues, but nearly every scientist featured was male. (A colleague brilliantly pointed out that Lucy and Ardi were the only females in the paleoanthropology special.) I felt deeply that not only was I being overlooked; my awesome female colleagues were also being overlooked. I also felt so very alone—I didn't have a community of people who looked like me.

Lexi was an aspiring film director and suggested: "How about if you wear a beard while doing fieldwork and I come out and film you?" Thus was born the Bearded Lady Project.

The goal of the Bearded Lady Project is to celebrate female paleontologists and their work. Working with fine arts photographer Kelsey Vance, Currano and Marsh produced portraits of more than one hundred women paleontologists wearing beards. The photos by Vance are "old-fashioned images that capture the lost legacy of women in paleontology—the black and white pictures of pioneering female scientists that are absent from our textbooks." At same time, Marsh was filming a documentary, *The Bearded Lady Project: Challenging the Face of Science*, that premiered in March 2017. The documentary examines the work of female paleontologists, who tell their own stories of fieldwork and the challenges they have faced. The goal was to address the lack of "good portrayals in film (fiction or nonfiction) of talented, ambitious women having each other's backs and helping each other to succeed." As Ellen describes it: "Through the project, we built a sense of community, a 'sisterhood of the traveling beards,' as Karen Chin put it."

Overall, then, there is notable progress in the status of women in paleontology, although clearly there is still a way to go. In a 2017 message to the membership, Arnold Miller, then president of the Paleontological Society, said:

There is a strong sense that the Society has not adequately supported and promoted the recognition and professional advancement of women; that it has not actively sought to enhance the representation and advancement of underrepresented minorities in the Society; that it has not been responsive to the harassment of women at meetings, in the field, and at other venues; that it has not devoted sufficient attention or resources to efforts intended to recruit underrepresented minorities or women to the science of paleontology; and that it has not addressed barriers encountered by our LGBTQ and disabled colleagues. The time is at hand to address these and related issues forthrightly, and to begin actively ameliorating them.

In the same message, Miller notes that underrepresented minorities make up less than 5 percent of the membership. Although all sciences have relatively few Hispanics, African Americans, Native Americans, or Asian Americans, this is especially true for the geosciences, including paleontology.[13] Geosciences are the least diverse of the STEM (Science, Technology, Engineering and Math) fields. This is true not only for currently employed geoscientists but for the students who are in "the pipeline." A 2010 study by the National Science Foundation showed that members of underrepresented minorities received about 16 percent of bachelor's degrees in STEM between 2000 and 2008, but only around 6 percent of degrees in geoscience.

The American Geosciences Institute, an umbrella organization for the numerous geoscience societies, estimated in 2017, based on exit surveys, that 75 percent of undergraduates in the Earth sciences identified as Caucasian, 3 percent as Asian, and 8 percent as "Underrepresented Minorities" which includes Blacks, Hispanic/Latinos, Native Americans/Alaskans, and Pacific Islanders. This last group was dominated by Hispanic/Latinos.[14] These numbers should not be a surprise to anyone who has attended a meeting of paleontologists; although minorities are becoming more visible, we are still very white. Much of the diversity comes from the attendance of our many foreign and foreign-born colleagues.

Why this is true, and the best way to remedy this situation, is something many of us are concerned about. Suzanne O'Connell and Anna Holmes suggested in 2011 that the major factors that attract students to the geosciences may be less effective for underrepresented minorities.[15] As I discussed earlier, the critical point on the career path for many paleontologists was an exciting college course or an inspiring professor in the geosciences. These courses and mentors in the discipline tend to be lacking at many historically Black universities or tribal colleges that minorities might attend. Another factor O'Connell and Holmes suggest is that love of the outdoors, including visiting national parks, is less ingrained within minority communities (my colleagues who belong to these communities vehemently disagree with this idea). A final factor on their list is a perception of lower prestige of the geosciences by family members, who may greatly influence student choices.

These ideas were examined for Hispanic students in a survey of University of Arizona students.[16] Although the lack of out of school outdoor experiences were important, their results suggested that familial factors were the major impact, with families not supporting the choice of major. Vertebrate paleontologist Bhart-Anjan Bhullar's parents came from India; they were fine with his career choice, but he knows that other Indian families did not think paleontology was a viable career—that it is was "not practical." In an interview, Lisa White suggested that "students of color may know scientists in their families or communities, and they're usually physicians, or maybe they're engineers. They're doing work that can be perceived as more valuable, more connected to people's well-being."

A different approach is taken by Jacqueline Huntoon and her colleagues.[17] They identify imposter syndrome and stereotype threat as two factors that impact minority students. Huntoon and her colleagues suggest that imposter syndrome is especially a problem with minorities because they have so few successful role models. Stereotype threat is the concept that because the group the individual belongs to (Hispanics, Blacks, women . . .) is believed to be bad at math or science, they believe that they must also be bad at these subjects. Dena Smith, a Latina paleontologist who is now a program director at NSF, needed to overcome just such an issue. She was told by high school teachers and by family members that women and minorities were bad at science and was intimidated about these subjects as a result.

Efforts to increase the representation of minorities[18] within the paleontology community and the geosciences in general are further behind than those to increase the number of women, although multiple initiatives are ongoing or are in the works. The National Science Foundation has funded multiple programs designed to increase the participation of underrepresented minorities in STEM disciplines. The Geological Society of America has created the On to the Future Program, whose main goal is to financial assist students from diverse backgrounds in attending their first GSA meeting and to provide access to information and mentoring while they are there. The Paleontological Society has formed a committee on Diversity and Inclusion and is exploring cooperation with SACNAS, the leading organization fostering Native American and Chicano/Hispanic success in STEM disciplines, and with GSA.

Key to any future changes will be education and outreach; we need to positively encourage *all* children who are interested in paleontology, but also make sure that future generations of every background are more comfortable with all of science. Dena Smith points out that people feel disconnected from science; they do not know that "science is theirs," that it is an inseparable part of their life and what they are doing. She suggests that this is partially the fault of scientists themselves; that we have been arrogant and disrespectful, sending the message to women and minorities that they "are not scientists." A great deal of recent emphasis in science education is to correct this misperception by providing activities where students actually do science ("authentic research") rather than just being told about it, which can empower them to seek careers in the field.

One issue that concerns both women and minority group paleontologists is that they are continuously being asked to serve on committees that need to fill their implicit diversity quota. Alycia Stigall related her experience: "Once I was asked to serve in a leadership position for a geo/paleo group because I was a dynamic woman from North America and balance was needed in gender and geography. No mention [was made] of any scientific qualifications I had to serve as a leader for the group. . . . It was extremely insulting and felt really demeaning to be considered a category instead of scientist." This "checking off the box" for a member of an underrepresented group poses a special conundrum for untenured faculty, who might feel the pressure to participate but really should not get distracted from efforts more likely to get them tenure.

For both women and minorities, having mentoring and positive role models are very important. As Phoebe Cohen told me: "Mentoring, informal socializing that leads to collaborations, and having senior colleagues advocate for you is a huge aspect of this in my opinion, especially of junior and mid-career faculty." Lydia Tarhan feels that her experiences as a woman paleontologist have been very positive, in large part because she was able to work for her PhD with Mary Droser at the University of California, Riverside. Kay Behrensmeyer was also fortunate in having the opportunity to interact with the famed paleoanthropologists of the Leakey family in East Africa: "Mary Leakey was a determined pioneer in leading scientific excavations at Olduvai Gorge, thus a wonderful role model for me and many other aspiring young women in field science, especially in African paleoanthropology." Although Pedro Marenco lacked Latinx role models who could help guide him through the challenges of getting through graduate school and into the professorate as an underrepresented minority, he now finds that he fills that niche for his own minority students. They are more comfortable interacting with him because of shared backgrounds. The same is true for Bhart-Anjan Bhullar, who finds that he attracts students to his lab that also come from immigrant families. He strongly believes that a diverse workplace is more compassionate and inviting and encourages different ways of thinking. Lisa White, a Black woman micro-paleontologist, is accustomed to playing the part of role model and being asked to reach out to youth, but she feels that paleontologists need to learn from other disciplines how to be better mentors. I witnessed the importance of role models firsthand when I saw the reaction of the Latina students in my department to an informal brown-bag with Dena Smith. Meeting a successful woman scientist from a similar background was very exciting to them.

Seeing minorities in positions of authority and being honored for their achievements is also important. Rowan Lockwood is currently education/ outreach coordinator for the Paleontological Society. She points out that it is obvious at the annual PS business meeting and reception that none of the officers or award winners is a person of color.

Events in the spring of 2020 brought all of these issues to the fore. Seeing images of a Black jogger run down and shot, a policeman murdering a Black man in plain sight, a Black woman killed by the police in her own apartment, and a white woman calling the police on a peaceful Black birder in the park galvanized much of the scientific community to address its own failings to address bias and racism. The Paleontological Society was on the leading edge of this, almost immediately issuing a powerful "Statement on Anti-Racism," which was prominently placed on the society's homepage. They also prepared a guide to resources for anti-racism allies.[19] The Society of Vertebrate Paleontology (SVP) also put out a strong statement on Facebook that acknowledged racist aspects of the field's history, such as removing fossils "from colonies without crediting indigenous discoverers" and that "SVP has very few Black members relative to our overall membership and that this low level of representation is reflected in the leadership of our society." SVP also formally endorsed the "Call to Action for an Anti-Racist Science Community from Geoscientists of Color: Listen, Act, Lead," which laid out detailed and specific proposals for action.[20] I am guardedly optimistic that the next decade will see true progress.

Paleontology has a small but highly active LGBTQ+ component. Schuchert Award winner Paul Koch publicly thanked his partner in 1998; I recall some surprise because this was a matter that was rarely spoken about openly, but mostly I remember thinking "good for him." Mark Uhen has hosted a gay and lesbian social event at the Society for Vertebrate Paleontology annual meeting for years. They are very well attended, including many graduate students and undergraduates. LGBTQ+ members of the Geological Society of America, including paleontologists, have been hosting a social event for over twenty years, and it has grown significantly recently as LGBTQ+ issues have become more mainstream and computer technology has allowed word to spread rapidly. In 2018, more than fifty LGBTQ+ geologists, from undergraduate and graduate students to seasoned professors, attended. Matthew Downen, a PhD candidate at the University of Kansas is conducting a survey of LGBTQ+ people in geology and paleontology. He found that "these socials are pretty crucial for younger paleontologists (and geologists in general) and contribute a lot to visibility and representation, whereas at their own schools they may not have visible LGBTQ+ people in their field." He also discovered that "paleontology seems to have a higher proportion of LGBTQ+ individuals compared to other subfields in geology."

Mark Uhen tells me that he is aware of no issues at meetings or negative impacts on careers. Michael Henehan, currently a postdoc in Germany, has also found that the field is accepting of him as a gay man. Nevertheless, formidable challenges remain. In an eloquent 2019 essay, the dinosaur paleontologist Riley Black described her difficulties as a transgender individual, including a lack of recognition and understanding within the science.[21] Riley discussed encountering homophobic ranchers during fieldwork and her reluctance to visit countries with a pattern of violence against queer people. As Riley's experience indicates, LGBTQ+ individuals are concerned for their safety in some situations and consequently some carefully choose where to do their fieldwork. Paul Koch agrees: "I have been out and accepted within the earth sciences and paleontology since I was a graduate student, but doing fieldwork in more conservative parts of the U.S. has had its challenges." Patrick Getty (Collin College) adds that "many also take into account local laws regarding LGBTQ+ issues when applying for and accepting jobs because it can be very difficult to live and work in a conservative region that is hostile to the LGBTQ+ community. As a general rule, members of this community seek out more liberal regions in which to live and work."

As is the case with most sciences, there has been little attention to accessibility, or the challenges posed by neurodiversity. Individuals with mobility issues will have difficulty in field settings and at conferences. Given the visual nature of most paleontological information, there are also hindrances to those with limited or no sight. These are not insurmountable; one of the most renowned paleontologists, Geerat Vermeij, has been blind since he was three years old. Vermeij has developed his sense of touch to study the shape and form of snails and clams. New technology has tremendous potential for overcoming barriers. The advent of inexpensive 3-D printing should make paleontological data accessible to a much wider audience, including those with limited vision. Virtual online field trips using Google Earth and other spatial databases will be a boon to those with accessibility issues, as will virtual conferences. The recently created International Association for Geoscience Diversity is devoted to "promoting access, accommodation, and inclusion" for current and future geoscientist with disabilities. They provide links to virtual and accessible field trips, resources for instructors, and a forum to discuss best practices. As the Miller letter states, however, the profession is only now starting to address the challenges stemming from ableist assumptions.

The topics discussed in this chapter were not easy to write about, especially from the perspective of a cisgender, straight, self-described OWM (older white male) who was the privileged beneficiary of a literal "old boys network." I do not have firsthand experience of many of the issues faced by my female, minority, LGBTQ+, and disabled colleagues, or the problems produced by intersectionality, nor am I privy to the discussions of things that they may be more comfortable talking about among themselves. The best I can do, and I hope my fellow OWMs do, is to listen with sympathetic and nonjudgmental ears and to be willing allies so that we can shift the culture of paleontology.

13

THE THIRD REVIEWER

There is no more trite an aphorism among scientists than "publish or perish." Unfortunately, it also has the ring of truth. As much effort as one might put into research or how clever one's ideas are, they are meaningless unless they are shared with your colleagues in the form of a peer-reviewed publication. And without a proven track record of publication, it is nigh impossible to receive grants to fund further research and to support students. Without grants and publications, the goal of attaining academic tenure and its guarantee of future employment becomes unattainable. Thus, if one does not publish, verily one will perish.

The kinds of papers paleontologists write fall into several overlapping classes. The most traditional one involves the description of a new fossil or perhaps the redescription of previously found ancient organisms based on new material. My first paper, where I was the second author with Niles Eldredge, bore the scintillating title "A Revision of the Pseudoniscine Merostome Genus *Cyamocephalus* Currie." To parse this, merostomes are a group of arthropods that includes the modern horseshoe crab; pseudoniscines are a division of the merostomes; and *Cyamocephalus* is a genus within the pseudoniscines, originally described by L. D. Currie in

1927 (names of species and genera are always in italics). We had new material that allowed us to write a new description. One can find papers like this in journals such as the *Journal of Paleontology*. Titles in a recent issue included the following:

- A New Species of *Coahomasuchus* (Archosauria, Aetosauria) from the Upper Triassic Pekin Formation, Deep River Basin, North Carolina
- Select Retiolitine Graptolites from the Aeronian and Lower Telychian (Llandovery, Silurian) of Arctic Canada

Stephen Jay Gould used to call these "A from B of C" papers; the title says what they are, how old they are, and where they are found.

Related to these are papers that describe whole new groups of fossils from a particular area and set of rocks or ones that review an entire group of organisms and put them in a broader context. Typical titles are the following:

- Select Stratigraphy, Mammalian Paleontology, Paleoecology, and Age Correlation of the Wasatch Formation, Fossil Butte National Monument, Wyoming
- A Combined Morphometric and Phylogenetic Revision of the Late Ordovician Brachiopod Genera *Eochonetes* and *Thaerodonta*

Papers like these are not generally accessible, even to other paleontologists. They are full of highly technical language and often are restricted to the group or area under study and familiar only to experts. There are highly specific requirements for the language used, and the quality of the illustrations and photographs must be superb. Although some of my colleagues seem to easily churn out a stream of these articles, I personally find them an uphill climb. They are also unlikely to receive wide attention unless the find is truly spectacular or unique. Nevertheless, these are in many ways the most critical papers that are produced. They are the bedrock of the field, comprising the fundamental data set on which everything else depends.

Another class of papers is of more recent vintage. These publications are often more theoretical and take a broader view of the fossil record. For example, my colleague Peter Wagner and I published an article that asked a simple question: What are the most common fossils? To do this, we used the Paleobiology Database (see chapter 7) and counted how many times a particular genus was found in a fossil collection. To our surprise, we found that only a tiny fraction of all genera accounted for a major fraction of all collected fossils. If you look at this paper, you will find lots of complicated graphs and tables but not a single photo of an actual fossil. Papers like these are the standard for the journal *Paleobiology*, where you might see titles like these:

- The Impact of Geographic Range, Sampling, Ecology, and Time on Extinction Risk in the Volatile Clade Graptoloida
- Ecomorphological Diversifications of Mesozoic Marine Reptiles: The Roles of Ecological Opportunity and Extinction

Again, these papers are often highly technical and require a sophisticated understanding of mathematics and statistics. As suggested by the journal title, *Paleobiology* publishes papers that focus on the biological aspects of fossils; that is, on treating fossils as the dead remains of once living creatures. These topics would not be out of place in biological journals that deal with ecology and evolution. Two representative papers could be:

- Select Polarity of Concavo-Convex Intervertebral Joints in the Necks and Tails of Sauropod Dinosaurs
- The Relationship Between Diet and Body Mass in Terrestrial Mammals

A third class of papers reflects paleontology's roots in the geological sciences. These papers focus on how fossils can inform us about ancient environments or how a knowledge of other aspects of geology, such as sedimentology or geochemistry, can improve our understanding of the biology of fossil organisms. For example, they may shed light on how a fossil organism became preserved. Papers like these appear in journals such as *PALAIOS* and have titles such as the following:

- Sediment Effects on the Preservation of Burgess Shale–Type Compression Fossils
- Distal Volcanic Ash Deposition as a Cause for Mass Kills of Marine Invertebrates During the Miocene in Northern Patagonia, Argentina

A final class are papers that focus on the environmental signal in fossils, such as interpreting ancient climates and oceans. These papers often contain, in addition to the fossils, detailed geochemical and geophysical data about the same sediments. These papers appear in many journals, such as the commercial journal *Palaeogeography, Palaeoclimatology, Palaeoecology*. For example:

- High-Resolution Isotopic Record of C4 Photosynthesis in a Miocene Grassland
- Extinction Patterns, $\delta^{18}O$ Trends, and Magnetostratigraphy from a Southern High-Latitude Cretaceous–Paleogene Section: Links with Deccan Volcanism

Of course, many paleontological publications don't fit easily into any of these classes and many straddle two or all three of them. There are also many articles that are not explicitly about fossils but that use paleontological data. And there are many more journals that publish paleontological papers. Some of these are specialist journals, such as the British publication *Palaeontology*, the international journal *Lethaia*, or *Marine Micropaleontology*. Many museums publish papers by members of their staff or by nonstaff who use their collections. Numerous national paleontology journals, such as *Acta Palaeontologica Polonica* (Poland) and *Bollettino della Società Paleontologica Italiana* (Italy), attract authors from around the world. Paleontological papers frequently appear in geology journals such as *Geology* or biology journals including *Evolution* or *Ecology Letters*. And then there are high-visibility publications such as *Nature, Science,* or the *Proceedings of the National Academy of Sciences* (PNAS). An unspoken practice among most scientists, not just paleontologists, is to submit a paper to one of these journals first because it is considered to be more prestigious, in particular, by various administrators. These articles also have a much higher probability of receiving coverage in the media. Unfortunately, these journals publish only a fraction of the papers submitted, so most papers end up in a specialist journal. This is not because they are categorized as substandard science; they simply may not be "sexy" enough to interest the editors of these publications. The choice of articles that appear in *Science* or *Nature* is the source of much quiet grumbling among scientists.

I should point out here that the venues for scientific publications are journals, not magazines! It took a long while to convince my parents that I was not paid to publish my research in a journal. In fact, authors often must find funds to pay the journal to support the publication of an article. However, this is one of the areas of academic science that is undergoing the most rapid change. I ought also to mention that many articles appear in edited book volumes rather than in periodicals.

The main goal of the scientific paper is communication, so the language needs to be as terse and clear as possible. In general, overly expressive language tends to be frowned upon. One paleontologist who could get away with literary flourishes was the late Stephen Jay Gould, who traded upon a strong liberal arts education. An important paper of his opened with "I have often imagined that La Giaconda's enigmatic smile is directed towards the scholars who write turgid books to explain her beauty. In length alone, literature about Shakespeare must exceed the Bard's works by an order of magnitude usually reserved for interstellar distances. Yet if every profession must have its *Drosophila*, shall we begrudge a Jurassic oyster yet one more article?"[1] In contrast, Gould's contemporary and friend David M. Raup was a veritable Hemingway. One of his best-known papers begins: "Many natural phenomena vary greatly in intensity and frequency. River discharge, wind velocity, and seismic activity are common examples. In each case, the lower intensities are most common and the higher intensities progressively less common. Events of greatest magnitude are generally rare and carry names such as 'flood,' 'hurricane,' and 'earthquake.' "[2] Raup's writing was vigorous, clear, and certainly not dull.

The format of the standard scientific paper, as it has been taught to generations of young scientists, is some variant of the following: Introduction, Methods, Results, Discussion, and Conclusions. The best-written papers tell a unified story. The Introduction sets the stage. In this section the writer justifies why this paper will be an important contribution to the scientific literature. It briefly surveys the state of the art in this area of knowledge, highlights some of the problems and questions, and indicates which of these the current paper will address. One of the most important parts of this is the citations: the careful noting of the work of others who have worked on the problem before. This does two things; it shows you have done the necessary scholarship before writing the paper and gives credit where credit is due. As you will see, the latter can be critical to a career.

Methods (sometimes Materials and Methods) is a bare bones recitation of what was done. The key thing here is that it theoretically enables others to replicate the work that was done. Unlike typical papers in biology or chemistry, paleontological papers often do not set out the details of an experimental protocol. Instead, they provide a detailed description of the geology of the locality that yielded the material, perhaps with a description of the collecting methods. This section may be called Geologic Setting. Alternatively, the sources of the data being analyzed may be reviewed, or this section may describe an experiment (yes, there are experiments in paleontology; see chapter 9). One recent trend I find annoying is to move Methods to the end of the paper as an appendix.

The next section, Results, describes what was found out because of the methods used. Again, there are a wide range of possibilities here that depend on the purpose of the research. The results can describe the outputs of analyses or the effects of the various experiments. When describing a new species, a highly technical review of the features of the extinct plant and animal are provided.

In Discussion, you say what it all means. Harking back to the Introduction, the results are put into the context of the questions raised there. How does the research in this study advance the state of the art? or What do we know now that we did not know before? For studies of previously unknown fossils, this section includes a detailed discussion of where they fit with previously described organisms.

All of this is tied up in the Conclusions, which serves as a summary of the overall paper. This is one place where you can wax a little bit lyrical; where you can try to sell the paper to the reader on how important, interesting, and, yes, downright cool the science is. The unspoken rule here is to end with some version of "more research needs to done." Tacked on at the end of all this are the References, which list all the papers consulted in preparing the article.

And, of course, every paper has a set of figures, sometimes line drawings, sometime photographs, and a set of tables. The importance of good figures cannot be overestimated. Today many figures can be downloaded as Power-Point files for teaching. One of the reasons the late Adolf Seilacher was so influential was due to the visual impact of his drawings, which often told a story.

Of course, like the so-called scientific method, this structure is an idealization and doesn't at all reflect how a paper is written or prepared for publication. The actual process is far messier and more idiosyncratic. For me, the process usually starts with a presentation at a national meeting. This has several advantages. First, it gives me a deadline to produce a coherent narrative. Second, it forces me to produce a set of legible and informative figures that tell my story. Third, it gives any coauthors I may have a chance to weigh in on what I am doing. Finally, it is an opportunity to get feedback on the basic ideas of the paper and to see if I am making sense.

As an example of the process and how frustrating and rewarding it can be, in 2014 I gave a talk at the Geological Society of America (GSA) meeting titled "Threatened Species in the Fossil Record: The Invisible Extinction." My coauthors on this were S. Kathleen Lyons (now at the University of Nebraska) and Felisa Smith (at the University of New Mexico). The research had started with a question I had: What would the current ongoing extinction (sometimes called the "Sixth Extinction") look like to a paleontologist in the far future? In other words, if we were to compare the extinction going on today with those in the past, we really should ask which extinct animals we would see in the fossil record. Most of the existing species threatened with extinction are vertebrates, such as mammals. My training, however, is in invertebrate paleontology, and I knew far less than I should to talk authoritatively about mammal extinctions. Fortunately, Kate and Felisa are among the world's experts on fossil mammals and on the fossil record of their extinctions. I was gratified when they agreed to work with me and make up for my deficiencies. They were able to supply data and insights that allowed me to tell my story. They also have a much better artistic sense than I do. A good figure should not only convey information but also be visually attractive and highlight the important parts of the message. Felisa and Kate made excellent suggestions on how to improve the figures that I used.

An aside on coauthors: If one goes back even thirty years, the vast majority of paleontological papers had a single or perhaps two authors. Following the trend in other sciences, multiauthored papers have become the rule rather than the exception today. Part of this is due to the laudable tendency to add students as coauthors. The other is the sheer growth in the complexity of the science itself, which requires multiple collaborators. A 2016 paper on the weird fossil *Tullimonstrum* had sixteen authors from three institutions. All of the authors contributed their own insights, often drawn from analyses on equipment for which they were responsible. Similarly, David Jablonski emphasizes "how important and valuable collaborations have been for me." Sometimes it takes a village to do science!

The GSA talk, in particular the figures, provided the framework for the paper we were writing. The next task was to decide where to submit the paper. This is not a trivial choice; each journal has its own strict policies on length and format. We decided that the paper was potentially of wide interest, so our first choice was *Science*. After several rounds among ourselves to tighten the language and improve the figures, we submitted the paper and anxiously awaited a decision. Within a few weeks we got the boilerplate response: "Because your manuscript was not given a high priority rating during the initial screening process, we have decided not to proceed to in-depth review. The overall view is that the scope and focus of your paper make it more appropriate for a more specialized journal. We are therefore notifying you so that you can seek publication elsewhere." Crap. A bit of explanation here. *Science* (as well as *Nature*) has a two-stage review process; the paper is looked at by members of editorial staff, who decide if it might be interesting enough to publish. Only then is it sent out for peer review (more on this soon).

Second try. *Nature* had interviewed me after my GSA talk and had even run a news article about it online, so they seemed like the obvious choice. This mostly involved some minor revisions to meet their style guidelines, a task made much easier by computer programs that allow almost instant reformatting of references. Off it went, and several weeks later: "Thank you for submitting your manuscript entitled 'The Invisible Extinction of Threatened Species,' for consideration. I regret that we are unable to publish it in *Nature*." Crap again!

Third try. We once more revised the paper, mostly fine tuning and reformatting, and submitted it to the *Proceedings of the National Academy of Sciences* (PNAS). The National Academy of Sciences is the most distinguished organization of scientists in the United States, and its members, who are selected in a confidential process, are the most eminent members of their professions. PNAS members act as the gatekeepers for the journal and decide whether papers should be sent out for further peer review. We were excited to get past this first hurdle, but then we got the following (you guessed it): "The expert who served as editor obtained 2 reviews, which are included below. After careful consideration, the editor decided that we cannot accept your manuscript." But there was this ray of hope: "However, because the reviewers think the work is of interest and the editor concurs, we are willing to consider one resubmission that constructively addresses all of the concerns raised in the critiques." A second chance! (I am told this is quite unusual.)

So, back to work. This is a short paper, so rewriting and resubmitting it was not too difficult. The key thing was to carefully read and address the issues raised by the reviewers (and to thank them for their efforts). Well, we *thought* we did: "After evaluating the re-reviews and your revised work, the editor has decided that it still falls short of the level of exceptional scientific quality and importance that we are seeking for publication in PNAS." This is about the time imposter syndrome kicks in—maybe I am not as smart as I think I am? By the way, two of the reviewers were quite positive about the paper, it was the dreaded "third reviewer" who hated it.

Fourth (or fifth) try. Once again, I am helped by my coauthors. They suggested we try *Ecology Letters*, one of the highest ranked journals in the field of ecology (journal rankings are based on the number of times papers in it are cited in other papers). Sure, why not? Revise, reformat, resubmit. We also changed the title to "The Fossil Record of the Sixth Extinction." And then we got: "the Referees make a number of comments with the aim of improving the manuscript. I invite you to revise and resubmit your contribution, and the subsequent publication decision will be based in part upon your point-by-point responses to the Referees' comments." Once more into the breach. And then in January 2016: "I am delighted to say that . . . your manuscript is now accepted for publication in *Ecology Letters*." Woo! Finally!

Of course, this was not the end of the process. Once a few minor issues were fixed, the journal turns the paper over to their editorial staff to look for remaining grammatical errors and screw-ups such as missing references (a bête noir of mine). The journal now sends out what are called proofs or galleys, which are a version of what will appear in its pages. This is the absolute last chance to catch that elusive typo. And then, finally, the sweet moment happens. The paper is published (first online and later in print). Time for the accolades to roll in!

For the vast majority of published papers, including mine, the result of all that work vanishes without a trace. Eventually we hope that colleagues will find it useful and cite it in their own work. In this case, however, I followed the advice that has been given to many of my colleagues: put out a press release and see if there is any interest in the media. I contacted my university press office, and we sent out a press release. To my delight, there was a great deal of media interest. Our press release was picked up by numerous science-related blogs and websites, as well as by domestic and foreign newspapers.

I was interviewed by a leading Spanish science magazine, by *Earth* magazine, and, ironically by a news writer for *Science*.

What I have described is not representative of the process for all papers, but it contains many of the elements familiar to my colleagues: the hopeful anticipation at the initial submission, the endless rewrites, the rejections, the frustration with the reviewers (especially the third one, who just didn't "get" what you are trying to say), and the joy of finally seeing your work in print, and the pleasure of sharing it with your colleagues and perhaps the wider world. What I haven't discussed are the stakes.

I recently retired as a tenured full professor. The chief personal value of publishing now is the satisfaction it brings. For my junior colleagues, especially those who are either graduate students or postdocs trying to land their first job or who have not received the sacred gift of tenure, the potential rewards and risks are much higher. For these scientists, the expectations for number of publications in high-ranked journals are considerable.

When I applied to my first job, I had two published papers. I had not yet published anything from my thesis. Today that number is considered par for the course for many undergraduates. It is not unusual to see graduate students applying for jobs with five to ten papers and postdocs with twice as many. Assistant professors are expected to be just as productive. It is not quantity alone that matters; the papers are expected to be published in high-ranking journals. And then there is the dreaded H-index and its ilk. Every time a paper is cited, the citation is entered into one of a few large databases. For research this is a great tool. If you are interested in a topic, you can easily find all the articles a paper cites and all the papers that cite it. It also, however, provides a quantitative way to assess the "quality" of a paper; it is assumed that the more important a paper is, the more times it will be cited. One of my papers, which of course I think is terrific, has been cited more than five hundred times. So the first measure of how good a scientist you are is having a highly cited paper. However, it is possible to have one highly cited paper with the rest remaining obscure. Enter the H-index. It is a simple metric: if I have five papers each cited five or more times, my H-index is 5. If I have ten papers cited ten or more times, my H-index is 10. As you can see, the higher your H-index, the harder it is to get to the next step.

The H-index is one of those tools that deans and committees love when deciding on tenure, especially when the job or tenure candidate is in a field with which they are not familiar. As you can guess, there are well-known tricks to game your number (for example, make sure your friends cite you, even if your paper is not germane). It also does a poor job of comparing among fields. There are far more ecologists and ecology journals than there are paleontologists and paleontology journals. It will be inevitable that the paleontologist will have a lower H-number than an ecologist, if viewed at the same state of their careers, with concomitant negative impacts on hiring and promotion. The fixation on these metrics can also have a deleterious effect on science; the incentives reward producing high-impact papers and discount the low-impact but vital papers that create the basic data of the discipline, for example, the systematic descriptions of fossil species or the detailed reports on a new locality.

Although as a writer I may grumble, one of the most important and somewhat thankless tasks of the professional scientist is the peer review. Scarcely a week goes by without an email requesting that I review a paper. For example: "Dear Prof. Plotnick, a manuscript entitled xxxx had been submitted to our journal. . . . Please let me know as soon as possible whether you can accept my invitation." Based on my own previous publications, the editors of the journal have decided that I am someone who knows enough about the subject of the paper to deliver my expert opinion on whether it should be published—I am "a peer." One must fight back the immediate desire to decline; after all, a properly done review of a paper can take hours or even days. Occasionally I am just too busy (sometimes with other reviews!) or feel the paper is not in my ballpark, so I decline. Often, however, I grit my teeth and agree. After all, my colleagues have taken time themselves to review my papers. I should pay it forward.

I usually work on the review in two stages. First, I read it over straight through. This gives me a feel for whether the paper is well written and meets the minimal scientific standards. If not, I can stop here and immediately recommend rejection. If it seems okay at this point, I then go over it much more slowly, making notes as I go along (I am old-fashioned enough to work off a paper copy), which are the basis of the review I will write and the recommendations I will make to the editor. I must emphasize that the goal here is not to be an SOB and look for reasons to suggest that the paper be rejected or to be rude to the authors. Instead, the goal is to identify places where the paper can be improved and to bring these to the attention of the authors. I always try to emphasize the positive aspects of what is there. It is rare to read a paper that is so amazing that it needs no changes (wonderful when it happens) and equally rare to find one that is so bad that it is irredeemable (depressing when it happens). Nearly all papers, my own included, are much improved by the comments and suggestions made by the reviewers. The recommendation "Accept with moderate changes" is par for the course. The hope is that the authors will resubmit to the same journal after they specifically respond to all the comments and describe how they were addressed. And, of course, they should thank the editor and all the reviewers for their efforts, even that nasty third reviewer.

Some last things about reviewing. Except for some confidential notes to the editor, the reviews are sent to the authors. The reviewer has the option of being identified to the authors or remaining anonymous. I usually choose the latter option, even if I write a positive review. Next, and I plead guilty to this, reviewers need to be timely. The review process is the one real bottleneck in the progress toward publication. Sometimes it is hard for an editor to find a compliant reviewer. Just as often, however, the review gets put on the back burner due to the demands of other work and months go by before it is addressed. For the young scientist, looking for his or her first job or hoping to get tenure, nothing could be more excruciating than waiting for the reviews of a paper. One very recent innovation is allowing the reviewer to invite a member of his or her research group (advanced PhD student, postdoc, or junior colleague) to participate in the review process. I have long wanted to do this; it is a great idea.

The practice of writing and publishing a scientific paper goes back centuries, to the early days of such august bodies as the Royal Society of London. By the middle of the twentieth century, the procedure was well established, with a standard series of steps between putting pen to paper (today it is finger to keyboard) and the production and distribution of the final product. The last decade, however, has thrown the entire process into an uproar, with a proliferation of pathways and fierce debates over the entire enterprise.

In the traditional model, the journal was published by a professional society, a museum, or a small number of commercial publishers. Museums generally publish papers by staff members, some of which can reach book length (monographs) and usually have a small distribution. Journals published by professional societies were supported by society members, who received subscriptions as part of their membership dues, and by institutional libraries. The subscription to the journal was a major incentive to be a society member, and most of us had shelves filled with scores of journal issues. Some society publications requested, but did not demand, "page charges" to help defray the cost of publication. Journals published by commercial publishers, such as *Palaeogeography, Palaeoclimatology, Palaeoecology* (aka *Palaeo-3*), were also available for personal and library subscription but were generally much more expensive than society journals. All of these were paper, often glossy, and of very high quality. Authors received paper copies of their paper, called reprints, which they could mail to interested colleagues. The income from the journals supported the costs of editing, typesetting, printing, and mailing. In sum, the costs of publishing were mainly supported by the subscribers.

This has all changed. As is the case with print newspapers and magazines, the advent of the internet has shaken the staid world of the printed journal and its publishers. The first change was online publishing. Journals began to offer online versions of their publications, originally as a supplement to the printed versions. Libraries were pressured to provide online access. Pretty soon the question arose of why anyone should pay to join a society to get a journal when the college library provided it online. And why should you fill precious office space with issues of journals, most of which you will not read except for one or two papers relevant to you? Societies were then faced with the challenge of providing other incentives to membership. At the same time, libraries began to recognize that they no longer had to provide shelf space for the thousands of journals to which they subscribed. That space could be freed up for computer workstations and study tables. Print subscriptions were canceled and back issues of journals went into storage or were tossed, to be replaced by scanned copies that were "just as good" (they are not!). Pretty soon online subscriptions became the norm and paper copies the additional cost option. Reprints basically disappeared, replaced by the ubiquitous pdf.

Most online only subscriptions were not any cheaper, however. Some journals, especially those of commercial publishers such as *Palaeo-3*, can cost a library thousands of dollars. Budgetary constraints then produce incentives to cancel subscriptions. Unfortunately, like cable television, commercial journals are often "bundled" into packages, making it difficult to cancel single publications. As a result, small, specialized publications, often of local interest (*Transactions of the Illinois State Academy of Science*) or published abroad

(*Acta Palaeontologica Polonica*), are the first to go. The descriptions of fossils or localities in them nearly vanish from sight.

There are advantages to online publication. I served on the editorial board of an online only journal called *Palaeontologia Electronica* (PE) that began in 1997. We offered our authors the ability to publish in full color (very expensive in print journals) and to post videos and other animations, including 3-D images that are impossible to provide on paper. Amazingly, PE was and is free to both authors and readers. It is produced on a shoestring and supported by small contributions from paleontological societies. It was open access before there was open access.

"Open access" is the other recent seismic shift in scientific publishing. As college and university libraries shifted to online only copies of journals, they also limited access to students, faculty, and others associated with the institution. A member of the public could no longer visit the library and read the journal there. The only other options were to subscribe themselves or to buy one-time access for the paper that interested them, which could cost $30 or $40 or more. Students and faculty at smaller, poorer institutions faced the same problem—they could not afford the subscription fee. Access to the free flow of information was being throttled by this pay wall. This struck many as unfair, especially if the research had been supported by government funding. Eventually, the major science funding agencies, such as the National Science Foundation (NSF), required that papers discussing publicly funded research be openly available a year after publication.

The solution was to shift the funding model. In open access publishing, the costs of publishing are now assigned to the author, who pays an upfront fee and can post the paper wherever the author chooses. The readers now have free and immediate access, no matter who they are. New journals, such as *PLoS One* (Public Library of Science) and *PeerJ*, sprang up that published exclusively online and were exclusively open access. These journals immediately siphoned off paleontological papers that previously had been submitted to more traditional journals. One advantage for authors is that press releases could include a link to the original paper, increasing its visibility.

All of these changes produced a crisis among established publishers, where the once firm ground was shifting rapidly. At the time, I was the treasurer of the Paleontological Society, which publishes the *Journal of Paleontology* and *Paleobiology*. The journals were both the major expense and the major income for the society. We needed to adapt and adapt rapidly. After considering several models, we switched to a structure that emphasized online publication but left open the option of paper copies for those who still wanted them. And we introduced two open access options. For authors willing and able to pay the fee, which we set to compete with journals such as *PLoS One*, we provided immediate online availability of the final published paper. This is called "gold open access." To meet the government mandate regarding publicly funded research for authors who could not pay the fee, a year after original publication these authors could make their final submitted version (but not the published typeset version) available at no cost. This is called "green open access."

After my term ended, it became clear to the leaders of the society that this solution was not sufficient to guarantee both the continued publication of the journals and the long-term survival of the century-old organization. I was brought back to head an interim committee to recommend long-term solutions. The end result is that the Paleontological Society got out of the publishing business. We turned over publication of the journals to Cambridge University Press, which makes an annual payment to the society to publish it on our behalf. We still technically own the journals but are no longer exposed to the financial risk posed by self-publishing.

I have mixed feelings about open access. On one hand, it is wonderful that science can appear online and be freely accessible so quickly. On the other hand, it creates a bind for those with limited funds to pay the gold open access fees, which can range up to $11,000 (seriously!).[3] My fear is that this creates a caste system, with the ability to publish high-visibility papers restricted to those with institutions willing to help out with the fees (mine will pay half) or fortunate enough to have grant funds sufficient to cover the costs. There is still much to recommend the subscriber pay model.

One final consequence of the move to online publishing has been an explosion in the number of journals. When journals were all printed, one went to the new journals area of the library and thumbed through the new issues to see if there was anything of interest. Some of my best ideas came from looking at journals from sciences other than paleontology to see if there was anything of interest. Those days are gone. Keeping up with the literature in one's own field, let alone areas more remote, is, as a well-worn phrase puts it, like "trying to drink from a fire hose." Surprisingly, one of the best online sources of information has been Facebook. My colleagues have been very good at posting links to new papers that they find exciting and that I otherwise would have missed.

I don't know what the future holds for paleontological publications. Much is tied up with the funding environment and the ability of authors to pay whatever fees the journal requires. There are suggestions to bypass formal peer review altogether; just post your article, and anyone can read it and post a comment, criticism, or suggestion. Rapid changes in technology will continue, and many of the format constraints may vanish. The ways scientific information will be made available over the next decades, let alone the next five years, is still unclear. Nevertheless, the ultimate goal—accurate and widespread transmission of new information—should remain our goal as we move forward into the future.

14

CONFERRING, CONVERSING, AND OTHERWISE HOBNOBBING

At some point during a large conference, I inevitably end up missing a colleague's talk and having to later stammer out an excuse, apology, rationale, etc. A number of years ago at a meeting I was handed a stick-on label with the simple phrase, "I'm Sorry I Missed Your Talk." Stealing this inspired idea as my own, for the past decade or so I have prepared a badge to wear at the annual meeting of the Geological Society of America (GSA). In big, bold letters it proclaims, "I'm Sorry I Missed Your Talk!!" This is followed by several other statements that I try to change every year. Some examples are:

- There was a competing session.
- I was out in the hall talking to someone important!
- At least it wasn't at the end of the last day.
- No one remembers these talks anyway!
- It took two hours to get lunch!

I would like to say these bring a big guffaw from those who read them. In reality, I must be satisfied with a small titter or a tiny smile. My colleagues share some of my frustrations in attending and presenting talks at a conference that might have seven thousand attend-ees and multiple competing sessions.

Scientific conferences are part of every paleontologist's year. They range from small specialist meetings that might attract two hundred people to huge generalist meetings such as the American Geophysical Union (AGU) that had twenty thousand attendees in 2016. These conferences have multiple purposes. First and foremost, they are a bazaar for the exchange of scientific information. Nearly all attendees take the opportunity to present some por-tion of their current research and to hear what others are doing. Conferences are where we first see cutting edge research that will appear in the next year or two in journals. Although a tiny number of talks and posters are dreadful, most are interesting and worth hearing. But what I really come for are those few talks that make me go "wow, that is really interesting!" It is especially exciting when the speaker is one of the younger scientists. They leave me reas-sured that our discipline will be in good hands and amazed at how far we have come in concepts, data, and methods. It is great to talk to these bright young paleontologists and to have my biases challenged. Paleontology is not an intellectual backwater!

Second, for many of us, this is the only opportunity during the year to meet face-to-face. We reconnect, both professionally and socially, with our many colleagues and friends who share our interest in the history of life. If you are working jointly on a project, it is a good time to sit down over a meal and review progress and plan future activities.

I have known some of these people for four decades. Others I met only a few years ago or am just meeting now. One thing is universal; we usu-ally greet each other with handshakes or hugs. There is a genuine sense of warmth to the atmosphere of the meeting. The closest parallel I can think of is a large family reunion. The core membership have known each other for years and can swap stories of how things used to be, and the new mem-bers are being introduced for the first time and made to feel at home. Much of the closeness also comes from the somewhat incestuous (figuratively) nature of our science. Paleontology is a relatively small discipline; many of us had the same PhD advisors, or our advisors had the same advisors, making us academic "cousins." We send each other our best students to continue their education. This can also, unfortunately, add an appearance of cliquishness.

Third, conferences often coincide with business meetings of professional societies. Because everyone is going to be at the meeting anyway, it is a cost-effective way to bring all the officers together in one place. For three years I served as treasurer of the Paleontology Society. A good deal of my time at conferences was spent in private and public meetings about the society's affairs.

Finally, it is a good time to pick up some swag. Publishers, societies, government agencies, universities, and equipment suppliers have booths open for much of the duration of the meetings. Usually, they have some small giveaways. Some are quite cool, like the geology hammer-shaped USB drive I got from the Geological Society of London. If you forgot to bring a pen, there are plenty to be had. You can also buy a recent book in your field at a significant discount from the regular price.

Preparing to go to a conference begins many months in advance. The 2016 annual meeting of the GSA began on September 25. The deadline for submitting an abstract for a presentation at the meeting was July 12. An abstract is a short summary, usually no more than four hundred words, of what you want to talk about. Given the time lag, it is sometimes hard to be too specific, although you can't simply say "cool stuff will be discussed." There is usually a small fee for submitting an abstract, and they are technically reviewed, although given the volume of submissions only the most egregious problems lead to rejection. For most meetings, you are restricted to giving one presentation, although you can be a coauthor on others.

A month or two later, you hear if the abstract has been accepted and when you are giving your talk. If you were clever, you reserved a hotel room as soon as they were announced by the organizers, so you are not stuck paying $300 per night. You also need to arrange flights to and from the meeting venue, which you hope is in a fun place to visit, with good and convenient bars and restaurants. Based on past experiences, many of us would avoid going to conferences in Houston, Reno, Atlanta, or Orlando. Vancouver, Seattle, Denver, and Portland are great. I once went to a specialist meeting in Barcelona. The meeting was boring as hell, but the venue was amazing.

The main task now is preparing your presentation. There are two options at this point: an oral presentation or a poster, usually decided by the meeting organizers. Each has its own strengths and weaknesses. Oral presentations mean standing up in front of an audience for about fifteen minutes (sometimes twenty) and giving a slide talk as part of a session. A session can run for three to four hours and include about fifteen talks. The advantages of the oral session are that you are "on stage" for only a short time, and you can be preparing up to almost the very last minute. If you are a good speaker with good slides and have interesting science, you can impress a lot of people all at once. There are number of downsides. There is almost no opportunity for feedback. Theoretically you are supposed to leave a few minutes for questions, but almost no one ever does. Given the length of the sessions and that several competing sessions are usually going on at the same time, it is also very easy to be overlooked.

To a large extent, timing is everything. The worse slot is the last talk on the last day; many people have already left or are too burned out to pay attention. Another bad spot is right after lunch because folks are often late getting back. The last talk before lunch is also a problem because people are rushing out to find a spot to eat nearby, as is an eight o'clock talk on later days of the meetings (out partying the night before). In 2016 I had an ideal time, the very first talk on the very first day, a time when my colleagues were at their peak. The rest of the meeting was worry free. I was also in the first session of the first day in 2017. In 2018, karma paid me back by putting me in the dreaded last session. Not as bad as I had feared.

An oral presentation is also a very bad place to be if you are subject to stage fright. A few of my colleagues, even those with years of experience, are comfortable only if they can read their talk from detailed notes. Many younger scientists incessantly rehearse their presentations. A student from another university that I worked with gave a talk at a regional meeting. You could see the terror in his eyes; the phrase "deer in the headlights" came to mind.

Poster sessions are quite different. A poster carries much of the same information as an oral presentation but is presented on a nicely printed sheet about 4 feet by 6 feet. The poster needs to be printed well in advance and transported to and from the meeting. You can always tell that someone on your flight is going to your conference if there is a large tube in their carry-on luggage. The posters are displayed as part of a "poster session," which can run the entire length of a meeting for small specialist conferences or half a day at large meetings. The presenter is expected to stand in front of the poster for at least an hour, sometimes more, to explain the content with interested attendees and to answer their questions.

The biggest advantage of a poster session is the opportunity for detailed feedback and discussion; it is far superior to an oral session in this regard.

One drawback is the inability to make last minute changes. It can also be exhausting to stand in the same place for hours and give the same explanation repeatedly. Worse than that, however, is having person after person walk by without looking at the result of your hard work. This is where eye-catching graphics and a good title excel. It is difficult to see a young scientist standing by a poster with a look of slight desperation and wondering why nobody is stopping to talk. I suspect speed dating and Tinder may not be dissimilar psychologically. I make a point of talking to as many students as I can; many of my colleagues do the same.

One recurrent issue is the status of posters. It is an implicit assumption that oral presentations have higher prestige. The majority of poster presentations at GSA are by students, often attending for the first time. For me, this is an opportunity to identify students who might want to work with me in the future. For more senior scientists, being given a poster slot can be viewed as an insult. A colleague of mine who oversaw the schedule of talks once apologized profusely for assigning me a poster session. I had to explain that in this case I had requested a poster because I wanted feedback at this stage of the research.

Going to all the talks and posters is impossible. There may be four competing oral sessions and a poster session going on at the same time. And there may be scores of posters. In addition, you might get into a conversation in the hall with colleagues and totally miss a talk you had planned to hear. I have finally given up feeling guilty about this. If the science is important, I assume I will see it in a paper in the next year. And I have my excuses badge.

Every conference has its own dynamic. Some of it is related to size and the duration of the meeting. The national GSA meeting may have seven thousand attendees, and it lasts for four full days. In addition to this, the Paleontological Society usually sponsors a free "short course" on a particular topic the day before the meetings begin. If I stay for the full conference, I am burned out by the last day. By that point, I have seen everyone I want to see and risk falling asleep during someone's talk. There is a reason one hopes to avoid speaking on the afternoon of the last day. Still, I always come back from these meetings totally excited about the science, full of new ideas for my own research, and happy to have seem my many friends and colleagues.

In addition to the large national meetings, the Geological Society of America also sponsors regional meetings that draw attendees from neighboring states, such as in the Midwest. These get-togethers are smaller and shorter, often only two full days. Many faculty and students from small colleges attend because they are closer and less expensive than the big national meetings. Again, it is a good place to look for prospective students. These conferences also have associated field trips, which I may be able to use in teaching my own students.

Then there are specialist meetings for those particularly interested in a narrowly defined area of research. For example, I am very interested in trace fossils, the remains of behavior preserved in things such as burrows and borings. Specialists in trace fossils (ichnologists) from around the world gather every four years at the International Congresses on Ichnology (Ichnia). The last two Ichnia meetings were held in Newfoundland and rural Portugal. These are wonderful meetings; it is the only chance I have to meet my colleagues from Spain, Brazil, China, Russia, Hungary, or elsewhere who simply cannot afford to come to the United States for a conference. Because the focus is on one topic, the opportunities to learn and to help advance the cutting edge of the field are great. And admittedly, it is great to have an excuse to travel abroad.

Choosing which of the many conferences to attend is where the cultural differences among paleontologists stand out. As an invertebrate paleontologist, my chief annual meeting is the Geological Society of America (GSA), which is also the annual meeting of the Paleontological Society (PS). Although the PS is an umbrella organization for all paleontologists, in reality it is dominated by invertebrate paleontologists and paleobiologists. Many presentations at the GSA are strongly grounded in geology. A vertebrate paleontologist may attend the GSA but most go to the conference of the Society of Vertebrate Paleontology (SVP). Similarly, workers on fossil protists often attend the American Geophysical Union (AGU) meetings, where there are many sessions on ancient oceans and climates. Paleobotanists frequently go to meetings of the Botanical Society of America. Paleontologists with a strong geological bent may attend the meetings of the Society for Sedimentary Geology (often referred to as SEPM). And, of course, paleontologists in other countries have their own national and regional meetings. Except for the infrequent North American Paleontological Convention, which has been held eleven times beginning in 1969, paleontologists rarely meet as a unified group. This disunity within the field has made it difficult to coordinate planning on research initiatives; to support cases to be made to the NSF, the wider scientific community, the public, or members of Congress; or to make agreements on controversial topics such as fossil collecting on federal land.

I have never attended an SVP meeting, but my colleagues who do go paint a different picture from what I am familiar with at the GSA. Scientifically, the SVP differs from the GSA in that many of the talks are much more biology oriented, with a strong emphasis on morphology (animal form) and the recently emerged field of evolutionary developmental biology (evo-devo) that merges the paleontological record of evolutionary change with genetic studies of how modern organisms develop. This meeting has also gotten quite large. Anne Weil (Oklahoma State University) thinks it is about five times bigger than when she started attending around 1990. There has been a concerted effort to get students to attend, and a large contingent of amateur paleontologists, mostly dinosaur enthusiasts, flock to the big names in the field. Despite the increase in size, the SVP has the same sense of family as the paleontology get-togethers at the GSA. Sara ElShafie, a graduate student at Berkeley, describes it as a collegial, friendly, and supportive community, with even the "big names" going out of their way to greet her and make her feel welcome. Along with the growth in size, there have been some changes in culture. When the meeting was smaller, the concluding banquet featured speakers, some of whom did not know when to stop talking. At some point, however, a food fight could break out! The after-dinner speakers are now reduced in duration, to be replaced by a DJ and dancing. Based on the videos I've seen of this celebration, I need to come up with an excuse to go to the SVP.

In the *Structure of Scientific Revolutions*, Thomas Kuhn gave one definition of *paradigm* as "the entire constellation of beliefs, values, techniques, and so on shared by the members of a given community." Members of these communities read the same journals and belong to the same societies. Attending a scientific conference is another way to demonstrate that you are a member of the community, and it is how students are indoctrinated into membership. Part of this is being welcomed by established members. For a young scientist, there is the excitement of associating the face of an attendee with the author of a paper you have read. If the senior scientist is particularly famous, it is an opportunity to be awestruck in the presence of greatness. The late Stephen Jay Gould always had capacity attendance at his talks because, well, he was Stephen Jay Gould. I usually get a blank stare or a more gratifying, "I read one of your papers." Most scientists, no matter how eminent or famous, are friendly and more than glad to spend some time chatting with enthusiastic young paleontologists.

For decades, one of the highlights for the young crowd was "being Dolfed." Adolf (Dolf) Seilacher was one of the giants of the field. When teaching paleontology, he is one scientist whose name comes up again and again. Dolf was a seminal figure in studies of fossil preservation (taphonomy), the behavior preserved in trace fossils (ichnology), and interpretation of the form of ancient organisms. Within the field of ichnology, Dolf was a demigod. He was physically imposing, very German, and very, very sure of himself. Though some of my colleagues have different definitions, to me being Dolfed meant describing to Seilacher what you were working on and then having him explaining to *you* that he was already familiar with that subject, had already figured it out, and that, by the way, you clearly did not read the German literature. At one GSA meeting, a session on trace fossils was held in his honor. Dolf was compelled to give his opinion on almost every talk. After my then student Karen Koy gave her presentation on experimental studies of modern traces, Dolf stood and said, "This is the future of the field!" OMG. Unfortunately, Karen had stepped out for a moment and missed the benediction.

Until recently, one of the veiled problems within paleontology and other scientific fields was sexual harassment (see chapter 11). This is particularly noticeable at conferences, where what Jess Theodor calls "cowboy behavior" occurs toward women, such as staring at breasts rather than the adjacent name tag, by a small but visible proportion of the men. My female colleagues have not only observed sexual harassment, especially of students, but experienced inappropriate handling and comments themselves. Discussing science was used as an excuse to ask more personal questions, including invitations for a drink or to a hotel room. Students subject to or witnessing this behavior may have decided that this is not the field for them. As in many other areas, there was a hesitancy to make an issue of the harassment because of the implicit threat to future careers.

With the increasing number of women scientists entering the field and assuming positions of responsibility, this part of the culture is now being directly confronted. The Society of Vertebrate Paleontology, the Paleontological Society, and the Geological Society of America have all adopted clear policies on sexual harassment, in the latter two it is part of a general Code of Conduct. For SVP, the initial impetus came from the female students themselves and was inspired by policies just put in place at Comic-Con. The SVP policy reads in part: "The society strongly disapproves of offensive or inappropriate sexual behavior at the annual conference and during any in person or virtual society functions and/or communications. . . . Any member who has a complaint of sexual harassment by another member should first clearly inform the harasser that his/her behavior is offensive or unwelcome and request that the behavior stop. If the behavior continues, the member must immediately bring the matter to the attention of the society's ethics committee . . . if the activity does not cease, the accused individual may be dismissed from the membership of the society." Both the SVP and PS statements very clearly state what constitutes sexual harassment.

There is also a more visible component to this. At the 2018 GSA meeting, many attendees wore a RISE button (Respectful Inclusive Scientific Events). Similarly, at the 2019 North American Paleontological Convention, individuals wore buttons reading "This IS PS" (Inclusive and Safe Paleontological Society). Anyone feeling harassed or observing someone being harassed was encouraged to report the incident to one of these individuals, whose responsibility was to listen and to pass on the report to the societies Ethics Committee, who could then act. Unfortunately, such action was needed at the NAPC.

With the advent of a younger generation of male scientists more sensitive to and better trained in these issues and of young women scientists with better tools to effectively respond to inappropriate behavior, I hope this part of the culture becomes a historical footnote. I am not immediately optimistic.

A related issue is serving beer or wine at the meetings or informal get-togethers at bars, which has long been part of the culture. In a 2019 blog post about AGU meetings, Laura Guertin pointed this out: "Want to be more inclusive? Stop making geology conferences about the beer."[1] In 2020, the GSA Committee on Diversity in the Geosciences recommended to the council of the society that "recognizing that the presence of alcohol during conference events can inhibit participation by some attendees and recognizing that speakers, presenters, and attendees must be present during talk and poster sessions, GSA should prohibit alcohol in the poster hall and oral sessions while technical sessions are occurring." Buried in a June 2020 news release was this item: "Council approved a recommendation from the Committee on Diversity in the Geosciences to prohibit alcohol being served in the oral and poster session areas at all GSA meetings and events."[2] Not everyone can, should, or wants to drink, and no one should feel pressured to do so. That said, I still would enjoy having a drink with my friends and colleagues when appropriate.

I wrote this prior to March 2020, and the Covid-19 pandemic has been an earthquake for nearly all aspects of professional and personal life. Scientific conferences have been no exception. Many have been canceled, others are playing wait and see, and still others, like many other activities, have gone totally online. I just participated in the online conference of the International Association for Landscape Ecology (IALE). How well did the virtual conference reflect the look, feel, and mission of the established in-person meeting? My feelings are mixed. There were real-time plenaries and symposia on Zoom, and as a vehicle for transmission of scientific information they worked well. In some ways, especially the ability to use various "chat" functions to leave questions and comments, they worked better than the standard format where there is often little or no time for discussion.

Instead of the usual poster session, there were "IPosters." They were presented in a basic poster format, but online presenters could add videos and audio narration as well as active links. As a medium for conveying scientific information, they were excellent. But they were not good vehicles for the social communication of science. There were no true face-to-face interactions, such as the serendipitous meeting of multiple people at a poster that leads to exciting discussions and is the meat and potatoes of a good poster session.

And, of course, a virtual meeting cannot adequately replicate the social interactions that make meetings so much fun. I could not run into old friends and colleagues nor easily make new ones. No hugs, no handshakes (although we may never do that again, even in a physical conference). I also missed the field trips, which have always been a highlight of IALE meetings.

Would I do another virtual conference? It depends. IALE was only three days long and was relatively small. That is about as many days of sitting in front of a screen as I can stand. It was cheap: no costs for flights or hotel rooms or meals out. But much of the fun of going to a meeting is to visit a new city and to have a nice meal with colleagues old and new. And nothing virtual can replace the intense chats in the hallway or the excitement of standing in front of a poster with a promising but nervous student. As most of us now know, zooming with your family does not replace being in the same room. The same is true for our academic families.

But there are also clear advantages to online conferences. In-person conferences are often difficult or expensive to attend, and some participants may be blocked from attending for nonscientific reasons, such as travel restrictions. Those with mobility issues may find conferences difficult to navigate. Many potential attendees may be concerned about the carbon footprint associated with air travel. Conferences also have the not inconsiderable expenses associated with registration, housing, and meals. There are also add-on charges associated with activities such as workshops and field trips. Total costs for a professional attending a large professional meeting, such as the Geological Society of America, can come to thousands of dollars. For those without institutional or grant support, money to attend a meeting comes from, as my colleague Tony Martin puts it, MYOP (my own pocket). Clearly there is a class bias based on finances.

But there are also other, often unconscious, class biases at conferences, which I share. I was prompted to think of them by two correspondents (at their request, I am anonymizing their comments). These mainly involve the social aspects of the meeting, which I wrote of so enthusiastically previously. I have come from a place of privilege in this regard: I attended prestigious institutions, my advisors were major figures in the field, and I had a large coterie of fellow graduate students. There was also a great deal of cross-pollination with similar institutions. The inevitable result is the formation of a community that is somewhat clannish and appears insular. It also favors extroverts. This creates real issues for those outside the community or those who are uncomfortable in social situations. A correspondent wrote to me, "I have found most in-person meetings alienating, exhausting, and lonely." Another correspondent echoed this, indicating that "conferences for me were always stressful, intensely isolating experiences, and while the content of particular talks was often interesting, the crush and noise of so many people ended up being overwhelming. I never knew what to say or who to say it to, and I never felt any sense of welcome or belonging. . . . For me that absence of community, that excruciating loneliness in a noisy throng, is what science feels like."

As a community, we need to recognize the authenticity of these experiences and find creative methods to make everyone feel welcome. This extends to others who might have barriers to participation, including those with disabilities[3] or those from disadvantaged areas or countries. The virtual meeting, as it evolves, is certainly a mechanism to address this. As one of my correspondents said, "I hope the advent of virtual meetings continues so I can hear from students and researchers in Africa, South America, Asia, Australia even, as loud and clear as North America and Europe, without shame around not being able to pay to go out for dinners." I encourage everyone to reach out to that student or colleague at a meeting who seems lost or alone. Let no one be alone in the noisy throng!

15

FIGHTING OVER SCRAPS

In the opening scene of Woody Allen's classic film *Annie Hall*, he tells an old joke about two elderly women sharing a meal at a resort dining room: "One of 'em says, 'Boy, the food at this place is really terrible.' The other one says, 'Yeah, I know; and such small portions.'" I think of this joke every time I discuss the funding situation for paleontology; it is really difficult to get funding, and if you do, the amount you get is limited.

Compared to high-energy physics, for example, where budgets can run into the tens of millions of dollars, paleontology is a very low-cost science. Except in remote areas, such as Antarctica, a summer of fieldwork can be done in the range of $5,000 to $20,000. Most of the cost is transportation and food, camping (when possible) is cheap, and supplies are usually off the shelf. If one needs to visit a museum to use microscopes or other necessary facilities, the major expenses are food and lodging. And paleontologists who analyze big data sets or produce theoretical models may need nothing more than a really good computer and fast internet. The expensive analytical instruments paleontologists use are usually shared with colleagues or are part of their institution's infrastructure. To be brief, paleontologists produce a lot of scientific bang for the buck.

The low research budgets of paleontologists are both a blessing and a curse. They are a blessing because we can continue to do our research with the bare minimum of support. They are a curse because there is an implicit bias that more expensive science is better and more important science. They are also a curse because decisions on hiring new faculty and granting tenure at many institutions are implicitly tied to the candidate's potential to attract big grants and thus income to the university through what is called "indirect costs" (overhead). One-third or more of a typical NSF grant is taken by the institution as overhead, ostensibly to pay for light, heat, secretarial services, etc. A large grant will generate correspondingly more overhead. For many universities, tuition, gifts, and endowment returns are not enough to fund operations; they have become increasingly reliant on external funding, especially federal dollars. This is even true of public universities, which have seen the percentage of their budget from state funding drop precipitously in the last twenty-five years.[1] At my own institution, unrestricted state revenues have dropped to about 12 percent of the total budget. As a former dean put it, we have gone from "state supported" to "state assisted" to "state located." The pressure on university administrators to increase revenue is transferred downward and affects faculty positions. As Steve Vogel wrote in 1998, there is an "institutional preference for expensive science,"[2] with deans favoring faculty that will generate more overhead. This preference has had a strong negative impact on fields such as paleontology that are not major revenue generators.

Although it is possible to draw on limited funds from within our institutions, which I have been fortunate in getting, most support for scientific research must come from outside. Unfortunately, the number of potential financial supporters for paleontological research is quite limited. Private and foundation funding is virtually nonexistent. The National Geographic Society and the Explorer's Club support some fieldwork, and the Petroleum Research Fund (PRF) of the American Chemical Society funds projects related to petroleum or alternative energy fields. Some funding, especially for students and early career researchers, is available from professional societies. Despite repeated attempts by the community, the fossil fuel industry has been very reluctant to directly fund paleontological research. This leaves government funding, in particular the National Science Foundation, although NASA has funded some projects, especially those dealing with the origins and early evolution of life.

The most visible unit within the NSF responsible for funding paleontology is the program in Sedimentary Geology and Paleobiology (SGP), which is part of the Division of Earth Sciences (EAR) within the Directorate of Geosciences. SGP funds paleontology as well as sedimentology and stratigraphy, so scientists in those fields all compete for the same funds. A 2006 study reported that the annual paleontology budget comprised 2.5 percent of the EAR budget and only 0.5 percent of NSF's overall geosciences budget. The SGP budget is currently about $7.5 million,[3] a number it has remained close to for a number of years.

A lot of this is not "new money"; much of it supports existing obligations informally called a "mortgage." On average, an NSF program has a mortgage of about 30 percent of its existing budget. In 2015, the SGP mortgage was a whopping 67 percent, leaving very little money to fund new initiatives (this percentage has dropped in recent years).

Some research dollars are available outside of SGP; paleontological research is also funded as a part of Arctic and Antarctic research, the Assembling the Tree of Life (TOL) program and other initiatives in systematic biology, and programs investigating aspects of Earth systems and global change. As of August 2019, there were 175 active NSF grants with at least some paleontology component, with a total budget of about $59 million. The largest grants are funded by multiple units within the NSF, represent large multiple investigator programs with a major infrastructure component, and are relatively extended in duration.

There are two or three opportunities (rounds) for funding within the SGP in any given year, although they have thankfully dropped strict deadlines for submittal of proposals.[4] In one of the most recent rounds, 126 proposals were submitted, with a total budget request of about $26 million. Of these, 18 proposals were funded, for a total of only $1.6 million. Another way to look at this is that although the probability of a successful proposal was about 15 percent, which is not terrible by NSF standards (but half of what it was 2001!), the average grant amount totaled less than $90,000.[5]

Ninety thousand dollars may sound like a substantial sum, but as Hope Jahren lays out in *Lab Girl*, only a fraction goes to doing science. Funding does not go directly to the scientists (the "principal investigators" or PIs) but to their institution. In many cases, at least a third goes to indirect costs. The PI and any other staff, including graduate students, must be paid, requiring additional grant funds to cover salary and fringe benefits. Tuition for graduate students is also usually charged by the university. Travel to present the research at meetings is usually included, as well as funding to support publication (open access papers can cost thousands of dollars to publish). Grants generally cover periods of two or three years, so these expenses are incurred each of these years. At the end, it does not leave much to do the proposed research.

The preparation submission of an NSF proposal is a long and complicated process that can occupy a large fraction of the working time of a scientist. A typical proposal is about forty pages long, with a strictly defined structure.[6] Collaborative proposals by scientists from multiple institutions can exceed one hundred pages. These are not trivial documents. Given the stakes, it requires careful attention not only to the science but also to the numerous technical requirements involved. Many institutions have full-time staff to aid in proposal preparation and have workshops to aid in the writing process. Numerous graduate programs have courses in proposal writing or incorporate it into other classes. As is so many times the case, coming from a well-funded institution confers a marked advantage in the ability to submit a high-quality proposal. A great deal of gamesmanship is also involved: How much should I ask for? Should I look for a hot topic or stick with one that may interest only me but that I know really well? How ambitious should the project be?

For many years, the proposal focused on the project itself—what is the scientific question to be addressed and how is this to be accomplished? Over time, the NSF has made it clear that the success of a proposal depends on both "intellectual merit" and "broader impacts." Specific criteria for these are supplied both to the submitting scientists and to the various groups that assess the proposal. The Intellectual Merit criterion encompasses the potential to "advance knowledge and understanding within its own field or across different fields." In other words, what is the broader scientific context of the proposed research; why should the results be of interest beyond the narrow confines of the project itself? Any good scientist should be able to easily address and satisfy this criterion.

Broader Impacts "encompasses the potential to benefit society and contribute to the achievement of specific, desired societal outcomes." Simply put, the project cannot be science for its own sake; it must show a societal benefit. What encompasses a broader impact is itself broad; it includes increasing the representation of traditionally underrepresented groups, improvements in STEM education, public outreach, commercial or national security applications, and public policy implications. The NSF is not restrictive in what constitutes a broader impact, but a discussion of it must be included in the proposal and cannot be given mere lip service. Incorporating well-thought-out broader impacts into a proposal has been a steep learning curve for many researchers, myself included.[7]

Like journal papers, NSF proposals are subject to a rigorous peer review process. The typical NSF process has two stages. The proposal is sent to anonymous ("ad hoc") reviewers, who are precleared to have no conflict of interest with the PI. As is the case with journal reviews, finding reviewers willing to give their time to this time-consuming effort is difficult (eliminating deadlines reduced the number of proposals and the pressure to find enough reviewers). The reviews are based on the NSF criteria, and the proposal is rated from Poor to Excellent. The proposals are then submitted to a panel of experts, who meet in person over several days to evaluate them (for proposals for graduate fellowships or postdocs, the meetings are virtual). Based on their discussions and often using their own reviews and those of the ad hoc reviewers, the proposals are ranked by the panel. These are recommendations only; the final decisions on funding are made by the professional staff of the NSF, the program managers. There are usually two program managers for each program. One of these is a permanent NSF employee. The other, dubbed a "rotator," is a member of the community who serves for a two- or three-year term. As Patricia Kelley once pointed out, by the time a rotator is broken in, he or she leaves.[8]

At the end of the process, whether funded or not, the applicant receives copies of the reviews, a summary of the panel discussions, and the comments of the program director. The whole process, from submittal to decision, can take as long as six months. For those who need a grant on record to be considered for tenure or promotion, the wait can be agonizing. (I speak to that from personal experience.)

I have been a recipient of NSF grants and have had many proposals rejected, sometimes for reasons I understood based on the reviews and comments, but just as often for reasons that left me puzzled. The program director is supposed to clarify the reasons and suggest grounds for improvement; my experience has been that they vary widely in their ability to do so. A rejected grant can be resubmitted after revision, but only after waiting a year. Some projects are finally funded after their third or even more resubmission. A grant that Robert Gastaldo requested had six PIs and a $900,000 budget for four years. The then NSF program director informed him that "if I fund you, I could not fund others!" It took six tries to get funded.

I have also been a reviewer and served on panels. I have been impressed by how seriously my fellow panelists take their responsibilities and how hard they try to be fair and honest in discussing the proposals. One major regret is that I did not take part in one of these years ago; seeing the decision-making process would have greatly helped my own grant writing. Ultimately "many are called but few are chosen." The number of worthy proposals far exceeds the funds available to support them.

The typical NSF proposal for paleontology supports at most a few investigators. It has long been believed by many, however, that the most effective way to increase the total pot of money for paleontological research is to steal a page from our colleagues in other fields: that is, develop large-scale projects that require the effort of numerous scientists, not only from paleontology but from other disciplines. This goal was the topic of numerous workshops in the United States and abroad and was long the mission of the late Harold R. (Rich) Lane, for many years the SGP program director.[9] Rich also emphasized the role of new technologies. The result has been a succession of ambitious programs with distinct acronyms, all of which began with great optimism and ended up producing relatively little of note, with some key exceptions.

The first of these was the Geobiology of Critical Intervals (GOCI). Chaired by Steve Stanley, then at Johns Hopkins, and sponsored by the Paleontological Society, GOCI was a 1997 "proposal for an initiative to support research in geobiology: the study of the coupled Earth-life system on geological scales of time," with the focus on intervals of geologic time that are "key episodes in the history of the Earth-life system."[10] The GOCI proposal was summarized in a detailed eighty-two page publication that emphasized how new techniques and multidisciplinary approaches can be applied to understanding these "critical intervals." Numerous examples are given, such as the origin of key plant groups, mass extinctions, and the early rise of animals. The document is an excellent summary of the state of paleontology at the end of the twentieth century. Unfortunately, the proposal was never formally acted on by the NSF. According to Steve Stanley, the then EAR division director was only interested in the last 120,000 years (since the last interglacial), while the second Iraq war's impact on the federal budget killed the chances of funding.

In 2006, Rich Lane was the SGP program director, and David Bottjer (University of Southern California) was president of the Paleontological Society. Bottjer recalls Lane insisting that "you can increase paleontology funding by putting together a big idea project with multiple people," but the "ideas have to come from the community." To address this, Bottjer organized a workshop on Future Research Directions in Paleontology (which I attended). Several initiatives were proposed at the workshop, one of which eventually became DETELON (Deep Time Earth Life Observatory Network).

DETELON was fleshed out at a 2010 workshop. The name mirrors another NSF initiative at that time, the National Ecological Observatory Network (NEON), comprised of more than eighty field sites designed to collect environmental and ecological data to understand how ecosystems are currently changing. The parallel mission of DETELON was to "address compelling, process-based questions about the interactions of environmental and biotic change in deep time that require the integration of multiple disciplines."[11] The DETELON proposal specifically called for "observatories" that would be funded for five to ten years and would involve not only paleontologists but a diversity of other geoscientists. DETELON itself was not funded. Lisa Park Boush, the SGP rotator at that time, stated that "conceptually it was a good idea, but the money was not there!" There was also collateral damage from numerous issues that the NSF was having with NEON, including management, budget, and goals, issues that are ongoing.[12]

The momentum from the DETELON initiative lived on, however. Yet another interdisciplinary workshop, led by Judy Parrish (University of Idaho), produced a report called TRANSITIONS. This became the Earth-Life Transitions (ELT) solicitation within the SGP, written by Park Boush. Announced in 2012, ELT was "a direct response to some of the grand challenges posed by the community. . . . ELT will support fundamental research into Earth system dynamics, focusing on scientific questions at the frontiers of climate change and biogeosciences."[13] The goal was to have three rounds of funding in 2013, 2015, and 2017, and to make awards of up to $1.5 million, much larger than the usual SGP awards. The first panel, held in the spring of 2013, funded eight interdisciplinary projects at about $4 million for five years. These included some of the proposed DETELON sites. This round was quite successful; the results were presented at a Paleontological Society Short Course and summarized in a book from the course.[14] But that was it. The second and third planned rounds of funding were never held. Part of the reason was due to changes in leadership at the SGP; Park Boush went to the University of Connecticut in 2013, and Lane died in 2015. NSF leadership once more changed goals: ELT was rolled back into regular solicitation and thus killed; it basically became invisible. It also left the SGP with the big mortgage mentioned previously.

Before he passed, however, Lane made one more effort. He had long wanted to bring together all the sciences represented by SGP—paleontologists, sedimentologists, and stratigraphers—to give direction and help increase available funding. Lane was also tired of having to fund workshops. Working together with key people from the Geological Society of America, the Paleontological Society, SEPM, and the petroleum industry, they put together a "community coordinating office" known as STEPPE (Sedimentary Geology, Time, Environment, Paleontology, Paleoclimate, and Energy) to "coordinate research, teaching, and learning in the areas of sedimentary geology and paleobiology."[15] The goal was to create a place where people could find each other and connect to collaborate: get them out of their disciplinary silos so they could develop large-scale projects and aid workforce development. There was also a heavy emphasis on developing associated cyber infrastructure and mathematical modeling. All of these were of interest not only to academics but to the industry. STEPPE was funded for three years beginning in 2013, principally by the NSF, with contributions from the participating societies. Judy Parrish came in as interim director and was succeeded by Dena Smith (box 15.1).

BOX 15.1

Dena Smith is an expert in the insect fossil record who is now at the National Science Foundation. Dena describes herself as a "dino-nut" kid, the only child in her California school that had a Land of the Lost *lunch bag. She also carried around a book about dinosaurs, which she was teased about because she was a girl. Dena always loved science and even at a young age was adamant that girls could do what they wanted. When she entered college at the University of*

California at Santa Cruz, she was an art major with the intent of specializing in science illustration. At UCSC, she was fortunate to spend a lot of time talking with the now retired Leo Laporte, who influenced numerous students who went on to careers in paleontology. Like Ellen Currano, Dena was fortunate to have a research internship program in the Bighorn Basin with Scott Wing. On her return, she changed her major to biology, eventually getting her PhD from the University of Arizona with Karl Flessa (an "academic father" to many successful scientists).

Like ETL, STEPPE was initially very successful. The office supported workshops, created a website designed to serve as a "central portal to information for the community," funded interns, and supported educational programs.[16] But like ETL, it was a one-off initiative and was not renewed. Those involved have different but not necessarily contradictory explanations: Lane passed away; NSF lost interest in supporting "offices" that key personnel had left; there were internal issues with finances; and the focus shifted away from the original goals.

Why have these efforts, each begun with great enthusiasm, careful thought, and extensive effort, failed to succeed? One component is the inability of many paleontologists to come together for a common vision. Karl Flessa (University of Arizona; NSF program director, 1988–1990) suggests that "we remain an individualistic, fragmented bunch. We can't even manage to regularly get the invert and vert paleontologists together at the same meetings. Consider how many little societies we have. We don't speak with one voice. That's what Rich was trying to overcome. He saw other disciplines solve that problem. We haven't." Howard Harper (SEPM executive director) agrees; the fields that STEPPE tried to bring together are diverse; people wanted to know "what's in it for me?" Likewise, Dave Bottjer feels scientists worried that a big project like DETELON would take money away from already scarce support rather than generate new funding streams. At the same time, the attempt to make a "big tent" also may have left the projects too unfocused. Dena Smith wonders, based on their track record, if "workshops/documents/big science on big questions are the best approach," although the workshops and documents are still needed to keep the NSF up to date on the status of the field.

Another issue identified by Smith is how the community sells itself. Patricia Kelley suggests that "it's in part an identity problem and battling the ingrained image of paleontologists as 'stamp collectors'"; that is, we are perceived as being interested only in collecting and identifying fossils. At the 2006 workshop, Lane explained that the poor NSF support for paleontology came not only from a lack of community-level initiatives but also from an image of it as an "old," not cutting edge, science.[17] This is a misconception with which paleontologists still struggle.

Clearly one issue is that the NSF is fickle; a lot depends on whether the types of science being done and the organizational structures are as fashionable at the time as the competition from other programs also asking for increased budgets. Another issue is reliance on one or two key personnel, who may leave at critical times. Programs in other fields that have succeeded did so because they had long-term, determined leadership.

Finally, there is the ongoing tension between technology and science in large programs. I was involved in an initiative called CHRONOS, whose goal was to put the diverse data developed independently by research in many areas of Earth history into a single integrated time frame and database. Unfortunately, the information technology aspects of the program became dominant. The scientific questions CHRONOS was designed to help answer were never clear, and as a result, the scientific community never bought in and the effort failed.

In marked contrast to CHRONOS is the Paleobiology Database (PBDB), which I discussed in chapter 7. In this case, there was strong community buy-in and a clear scientific issue, the history of biodiversity, to drive the research. As Lisa Park Boush has pointed out to me, despite being underfunded, the PBDB succeeded because it had a big vision that was well articulated.[18]

Another ongoing success with a strong paleontology component is the Integrated Digitized Biocollections (iDigBio), a central portal for information on some 125 million specimens housed in U.S. museum collections, including fossils. Some 40 million of these records include media. Many of these museum collections have never been studied. A recent paper found that about twenty-five times as much information on fossil localities may be available from museum records than is published in the PBDB.[19] Related programs include the Fossil Insect Collaborative, which digitized all major collections of fossil insects, and iDigFossils, which uses 3-D digitized fossil specimens for STEM education. Again, as Park Boush indicates, iDigBio has succeeded because of strong community involvement and by asking "What science questions do we want to answer" first?

The underlying problem remains the low level of available support, but one source of funding for paleontology remains untapped. It would be great to transform a tiny fraction of the public spending on fossils, dinosaur toys, and media into money for the science. Someone just spent $31 million to buy a *T. rex*. The movie *Jurassic World* cost $150 million to make and earned $1.672 billion worldwide. The entire *Jurassic Park* franchise has earned more than $5 billion. The Jurassic Foundation, funded by Amblin Entertainment and Universal Studios, funds research by early career dinosaur paleontologists and is an important funding source for them. The foundation gives out about $20,000 each year in grants, with nearly $1 million given out since 1998. Not to bite the hand that feeds us, but the Paleontological Society funds over twice that amount each year to all early career paleontologists, so I am a bit underwhelmed.

IV

DEEP TIME AND THE BROAD WORLD

16

THIS LAND IS YOUR LAND, YOUR FOSSIL IS MY FOSSIL

$31.5 million. That's how much a private buyer paid in late 2020 for a nearly complete skeleton of a *Tyrannosaurus* (STAN™) sold at auction in New York.[1] My first reaction was, "That's enough money to fund nearly all paleontology research in the United States for years!" My second reaction, one that I share with my colleagues, is that fossils like this belong in a public museum. As David Polly (Indiana University), then president of the Society of Vertebrate Paleontology (SVP), told *Nature* when commenting on a previous sale, "Fossil specimens that are sold into private hands are lost to science."[2] Unfortunately, this sale is part of a trend of the private sale of fossils at prices well beyond the means that public museums can match. This sale and others like it provide fresh fire to the long-running dispute between professional paleontologists and commercial collectors and to dissension within the paleontological community. The crux of the issue is whether fossils can be bought and sold as a commodity, or should all be treated as irreplaceable scientific resources that must not have a price put on them. It bleeds over to controversies involving private ownership of fossils and access to fossil sites on public lands. There are strong opinions and often bad feelings on all sides of these issues, with many of us feeling caught in the middle.

Although the buying and selling of fossils has a long history (after all, Mary Anning survived by selling what she collected), the modern fossil "gold rush" was triggered by what could be called the "SUE effect." In late 1997 the Field Museum purchased at auction the dinosaur known as SUE,[3] the largest and most complete *T. rex* ever found, for the then astonishing amount of $7.6 million (plus fees). It was by far the largest amount ever paid for any fossil, let alone a dinosaur. When I was interviewed about it by a reporter from a local television station, she asked if the large amount of money being spent was going to be a problem. My flippant response was, "No, it's not the museum's money. McDonald's and Disney are paying for it." In the short term I was correct that it was not an issue for the Field Museum, but in the long term I was woefully wrong.

The amount of money spent for SUE inspired the belief that *all* fossils are objects of potentially great financial value. This belief has thrown a shadow over paleontology, causing internal divisions within the professional paleontological community and damaging its relationships with commercial fossil collectors, landowners, and, to a lesser extent, with amateurs. This damage was exacerbated by the contretemps over the ownership of SUE, with competing claims from the private Black Hills Institute (BHI), which did the excavation and much of the original preparation; from the rancher who gave permission for collection of the fossil on what he considered his land; from the Sioux tribe on whose Cheyenne River reservation the ranch was located; and from the federal government, which held the land in trust for the tribe. At one point, the bones were seized by the federal government from the BHI and put in storage. An investigation of the institute led to criminal charges against it and its principals on charges not directly related to the dinosaur, resulting in jail time for its cofounder Pete Larson. The involved story of the collection and sale of SUE, the legal battles over its ownership, and the case against the BHI and Larson have been the subject of several books, a *Nova* episode, and a controversial 2013 documentary film. Two decades later, these impacts and bad feelings linger.[4]

First, a definition: a commercial collector is someone who derives most of his or her income from selling fossils. In some but not all cases, these are fossils they have collected themselves, thus the informal term "fossil hunters."

They range from single, often self-trained individuals to large and sophisticated operations like BHI that also have the expertise to properly prepare and display fossils. Depending on the fossils and the nature of their business, their customers can range from high-end buyers who want to put a dinosaur skull in their living room, to institutions hoping to add to their teaching collections, to amateur enthusiasts trying to fill out their collections with specimens they are not able to collect themselves. Specimens can be sold or traded online, at auctions, or at large rock and fossil shows, such as the one in Tucson, Arizona. They can also end up in local rock and gift shops, and to be fair, many have ended up in museums and research collections.

For purpose of full disclosure, I have purchased invertebrate fossils and am currently working on a project where I can obtain useable specimens at a local flea market for a few dollars. I have also spent slightly larger amounts at a fossil and mineral show run by my local amateur club to add a few common desired items to my own collections. And I have bought fossils to supplement my department's teaching collection, which largely consists of specimens purchased in the 1960s from large commercial suppliers. Nothing rare, and nothing of real scientific value.

Besides SUE, other cause célèbres in recent history, all involving vertebrate fossils, have damaged the already tense relationships between commercial collectors and academic (museum and university) paleontologists. In 2012, Heritage Auctions in Manhattan placed on auction the 75 percent complete skeleton of the large tyrannosaurid dinosaur, *Tarbosaurus bataar*, from "Central Asia." Before the auction could be held, a group of American and Mongolian paleontologists informed the government of Mongolia that the fossil without doubt came from there. Under Mongolian law, fossils are part of the nation's cultural heritage and cannot be exported; the implication was that the dinosaur had been looted and that its sale was illegal. The auction took place despite a judge's order obtained by a lawyer for the Mongolian government, with the dinosaur selling for about $1 million. The auction house decided to suspend the sale pending a legal resolution of the ownership of the fossil. It was soon discovered that the fossil had been in the custody of a Florida-based commercial collector, Eric Prokopi, and had certainly been looted from Mongolia and illegally imported. After an investigation by the federal government, Prokopi was arrested for smuggling, to which he eventually pleaded guilty in December 2012.[5] He served three months in jail. The dinosaur was returned to Mongolia, where it has been the inspiration for national interest in its paleontological heritage, led by Bolortsetseg "Bolor" Minjin, a Mongolian paleontologist who was deeply involved in stopping the original auction.[6] The recovery also shed light on the continued looting of Mongolian fossil sites for sale abroad. Other pilfered Mongolian fossils have been returned, including a skull of *Tarbosaurus* owned by the actor Nicolas Cage. Mongolia is not alone in this; Brazil and other countries have also been plagued by smuggling (see chapter 21).

In May 2009, a press release announced that "after two years of research a team of world-renowned scientists will announce their findings . . . the most significant scientific discovery of recent times." The "momentous find" was an exquisitely preserved early primate fossil formally named *Darwinius masillae*, informally named Ida, suggested to be ancestral to the higher primates. Behind the press release, in addition to the scientists and the museum involved, was a major media company. The gorgeous 47-million-year-old fossil came from the famous Messel fossil site in Germany, an ancient lake that beautifully preserved in shale a wide variety of organisms, including one of the oldest known bats. In 1983 a private collector discovered Ida. Like many other fossils, Ida was found by splitting between the laminations of a piece of shale, with some of the remains on one side of the split piece and some on the other side. In this case, one side preserved far more of the animal than the other. The incomplete sample was sold in 1991 to the private nonprofit Wyoming Dinosaur Center. Now it gets weird. At some point, an expert forger had added details to make this fossil appear far more complete than it really was. This led to an erroneous description of the specimen. In 2007, working with a fossil dealer, the owner sold the other, far more complete slab to the Natural History Museum of the University of Oslo for a reported $750,000. This was the specimen hyped in the media and described in the scientific paper.[7] The hype and the selling of the fossil raised hackles, not only among scientists but also among science journalists.[8] In a letter to the journal *Nature*, a group of distinguished primatologists stated: "Such objectionable pricing and publicity can only increase the difficulty of scientific collecting by encouraging the commercial exploitation of sites and the disappearance of fossils into private collections. . . . *We strongly believe that fossils should not have any commercial value*" (emphasis mine).[9] Riley Black (writing as Brian Switek) has written a highly readable account of the entire episode, including a review of the scientific status of the fossil.[10]

A final example is the "dueling dinosaurs," which has been called "one of the most remarkable fossil discoveries ever made."[11] The apparently intertwined skeletons of a Late Cretaceous *T. rex* and a *Triceratops* were found in 2006 by a rancher and fossil hunter, Clayton Phipps, on Montana ranchland owned by Mary Ann and Lige Murray. In 2013, the fossils were put up for auction but did not meet the reserve price of $6 million. In the years since then, the fossils became embroiled in the arcana of land rights in the American West. Land rights in some states are separable into surface rights and mineral rights, with the latter encompassing subsurface oil, gas, and minerals. When the Murrays bought the land, the previous owners retained part of the mineral rights; on this basis, they asserted that fossils were "minerals" and sued for a share of the proceeds from the sale of the specimens. In May 2020, the Montana Supreme Court ruled on behalf of the Murrays, saying that "dinosaur fossils . . . are not minerals under the word's common and ordinary meaning." This ruling determined who would get the proceeds of the sale; it also clarified future issues of ownership for both commercial collectors and professional paleontologists. On November 17, 2020, it was announced with much fanfare that the "dueling dinosaurs" had been purchased for $6 million by the nonprofit Friends of the North Carolina Museum of Natural Sciences and donated to the museum in Raleigh.[12] Additional millions will be sent to build an extension to the museum where visitors can see the skeletons being freed from the encasing matrix.

These sales raise the underlying issue of what a fossil, any fossil, is worth. Fossils occupy a peculiar position in our ability to valuate them. On one hand, they are items of undoubted scientific interest and thus have intrinsic value dependent on their value to science. On the other hand, they are also fundamentally interesting and attractive. Like shells and butterflies, they are collected and displayed as much for their intrinsic beauty as for their scientific interest. I can understand wanting to own a fossil for its aesthetic value; several pieces I've collected are on display in my own house.

The real problem comes from the same urge that occurs in every hobby that is based on collecting objects (I speak as a toy train collector). That is the desire to own the rarest or most unusual items. Common stamps, coins, books, or toy trains are not desirable and thus are sold at a low prices, whereas the rare or even unique items fetch a premium price. In addition, there are bragging rights to fellow aficionados of being the owner of this precious object. Unfortunately, although the value of a rare stamp is only what someone is willing to pay for it, the rarest natural history objects, such as some fossils, are also the ones with the greatest scientific value. I have a box full of the ubiquitous brachiopod *Composita*, which I give to children for free. This common fossil has educational value for the children but doesn't have a commercial value because it is common. In contrast, what makes the dinosaur sold in New York so expensive is not its intrinsic scientific value or the real cost of excavating and preparing it but the desire of its new owner (and the competing bidders) to say, "Look what I bought!" David Polly expressed a similar idea to me: "In my opinion, the root of conflict between science and commerce isn't sales and making a living, it is about equating fossils with art and auctioning them."

Yet another element of this is the idea, common to every speculative bubble since the tulip mania in seventeenth-century Holland, that buying and selling a rare object is also a way to make money. As shown in a recent series on the Discovery Channel, making money certainly is an incentive for many commercial collectors. It has also made collecting by professionals that much more difficult because landowners are becoming savvy to the possible value of the fossils on their property and are unwilling to grant access.

Because so many fossils are collected by commercial collectors and amateur paleontologists, many scientifically valuable specimens end up in private hands rather than in publicly accessible permanent archives. According to Lance Grande, other bidders for SUE included a real estate baron, a casino, and someone who wanted to display it in their living room! The key issue is that a private owner can block or limit access to critical specimens, making restudy impossible.[13] This issue came to the fore in recent years when a remarkable snake fossil with four legs, *Tetrapodophis*, was described from the Cretaceous of Brazil.[14] At the time, the specimen was on temporary display at a museum in Germany but was privately owned. The fossil soon met with controversy: first, because doubts were cast on its interpretation as a snake; second, because its original locality could not be established; and third, because it was back in private hands.[15] The Cambridge paleontologist Jason Head was quoted in *Science* as saying: "It's not repeatable, it's not testable. If any good can come out of *Tetrapodophis*, it's the recognition that we have got to maintain scientific standards when it comes to fossils . . . they have to be accessible." In 2020, the Society of Vertebrate Paleontology (SVP) specifically addressed the issue in a letter to journal editors: "privately owned fossils regrettably cannot be regarded as reliably available for study, cannot be considered part of reproducible science, and must not be introduced in scientific literature due to the uncertainty in the long-term accessibility necessary for guaranteeing reproducibility of data from them."[16] They urged editors of paleontology journals worldwide not to publish any papers describing fossils that are not "formally accessioned into a permanent, accessible repository."

To some, but far from all, paleontologists any sale of fossils is anathema. I know some vertebrate paleontologists who boycott the Geological Society of America meetings because they allow vendors who sell fossils. This opposition to commercial collecting is enshrined in the Society of Vertebrate Paleontology bylaws: "the barter, sale, or purchase of scientifically significant vertebrate fossils is not condoned, unless it brings them into, or keeps them within, a public trust. Any other trade or commerce in scientifically significant vertebrate fossils is inconsistent with the foregoing, in that it deprives both the public and professionals of important specimens, which are part of our natural heritage."[17]

The key phrase here is "scientifically significant." This may for most equate with rarity or newness, but to some it may not. In a 2014 essay by vertebrate paleontologists Kenshu Shimada, Philip J. Currie, Eric Scott, and Stuart S. Sumida, they insisted that "even the most common forms of fossils, such as trilobites and shark teeth, may be scientifically significant depending on the investigative questions asked. . . . Based upon this, if a fossil provides useful scientific data, then it is scientifically significant."[18] By extension, this implies that not even common fossils should be treated "as nothing more than commodities" because they potentially could yield useful data. They embed this in the larger idea that the public has been misled to believe that it is okay to buy and sell fossils, and it is therefore critical to reduce "through public education the demand of fossils," suggesting perhaps a replacement through paleontological art (paleoart). They also inveigh against private collections, saying that they do not meet essential standards of "permanency and accessibility." Shimada and his colleagues then lay down the gauntlet: "We therefore consider the battle against heightened commercialization of fossils to be the greatest challenge to paleontology of the 21st century."

I solicited this essay in my then role as commentaries editor for the online journal *Palaeontologia Electronica*[19] as part of a general call for commentaries on fossil collecting, spurred by the 2009 passage of the Fossil Resources Protection Act (more on this later).[20] The response from the commercial collectors came in separate commentaries, respectively first authored by Peter Larson and Neal Larson of the Black Hills Institute;[21] the former appearing in *Palaeontologia Electronica*[22] and the latter in a journal of the commercial collecting community.[23] Although the two responses differ in some details, they make many of the same points. First, there is a long history of professionals and the science benefiting from commercial collecting. Second, it is mainly in the United States that the two communities are so much at odds; in Germany, the UK, and Morocco there is far more amity and cooperation. Third, fossils that are left uncollected will be destroyed by erosion, and there are not enough academic paleontologists available to preserve them.[24] Fourth, commercial collectors are mischaracterized; although "commercial paleontology is not without its troublemakers," "the demonization and marginalization of a specific portion of the paleontological community is the result of misunderstanding, misplaced entitlement and simple intolerance." There is a code of ethics for the Association of Applied Paleontological Sciences, the professional organization of the commercial collectors.[25] Those who violate the code can be removed from the association. Finally, both essays urge "finding a way for amateurs, commercial fossil dealers and academic paleontologists to work together and do what is best for the public and the fossils."

All of this sounds good but does not address the key issues: the removal of important specimens into private hands; the loss of access to important sites for academic paleontologists; the ongoing gold rush for fossils incentivizing fossil removal by poorly or untrained individuals; and poaching and pillaging from established sites, public lands, or developing nations such as Mongolia. There also have been a proliferation of fakes and forgeries, as was the case with Ida and a "feathered dinosaur" from China dubbed *Archaeoraptor*.[26] As archaeologists can attest, laws and codes of ethics do not prevent illegal and unethical behavior.

Not all relationships between commercial collectors and academic paleontologists are fraught. Probably the most common vertebrate fossils in the world are the gorgeous fish from the Fossil Butte Member (FBM) of the Green River Formation of Wyoming; there are about ten commercial operations quarrying for these fossils. In his book *Curators*, Lance Grande of the Field Museum described how he built a mutually beneficial relationship with amateur and commercial collectors of Green River fossil fish, who "would notify me when something new or unusual was excavated. . . . The cooperation I have received over the years from the commercial quarries and amateur collectors in the FBM has aided my work immensely."[27] Grande points out that it would have been impossible for him to personally excavate as many fossils as the commercial operations uncover, by at least two orders of magnitude.

David Polly is similarly nuanced when it comes to commercial collecting, saying that many paleontologists, "including myself, recognize that commercial collecting that helps bring fossils into the public trust is both valuable and necessary. . . . Some private companies have a business model in which they excavate fossils, mold and cast them, put the original in a public trust repository, and trade in the casts (which they can sell again and again, unlike original fossils). In my opinion, this is good for everyone."

The amateur paleontologist John Catalani also firmly responded to Shimada and colleagues' critique, especially their comments regarding personal collections: "Although these collections are 'private,' they often meet the same 'essential standards' used in public repositories. . . . Serious amateurs understand the importance of documenting fossil specimens collected, and most amateur clubs have programs outlining this practice."[28] Bruce and René Lauer are models of avocational paleontologists who have gone out of their way to work with professionals and address their concerns. They have established the Lauer Foundation for Paleontology, Science, and Education[29] as the corporate owner of their well-curated collection that includes what is certainly the finest collection in North America from the Jurassic Solnhofen deposits of Germany (source of *Archaeopteryx*). The Lauers welcome professionals to visit their collections (I have had the privilege) and use their state of the art photography equipment. The foundation has established a succession plan that identifies a museum that will receive the collection.

The issues posed by commercial collecting cannot be separated from those involving access to collecting sites, especially those on public lands. Unscrupulous or unthinking collecting by individuals has long vexed those responsible for protecting our natural resources. One of the felony charges that stuck against BHI was for theft from Badlands National Park. Stealing fossils from public lands has long been a concern of the National Park Service paleontologist Vincent Santucci. He and John Ghist tell the sad story of the Fossil Cycad National Monument, an illustrative story of the "challenges associated with the management and protection of non-renewable paleontological resources."[30] Abundant fossil "cycads" (actually cycadeoids) from the Cretaceous were known from a site in western South Dakota by the early 1890s. Over the next thirty years, local ranchers and fossil dealers, as well as professional paleobotanists, removed a large number of specimens. To save the site, Fossil Cycad National Monument was established in 1922, but there was very little management or development by the government. By 1929, no visible specimens were found on the surface, although excavations suggested than many cycads remained buried. With nothing notable for visitors to see, Congress abolished the monument in 1957. Santucci and Ghist conclude: "Perhaps the lessons learned may be used to deter visitors to places like Petrified Forest National Park from engaging in souvenir hunting of petrified wood."

Regulations controlling fossil collecting on federal lands have long existed, but no unified approach was in use. In 2009, after many years of concerted lobbying by the Committee on Government Affairs of the Society of Vertebrate Paleontology, the Paleontological Resources Preservation Act (PRPA) finally became law in the United States.[31] This law provides strong new protections for scientifically valuable fossils, especially vertebrate fossils, on U.S. federal land, with the underlying concept that "paleontological resources are nonrenewable and are an irreplaceable part of America's natural heritage." This includes the complete exclusion of commercial collecting, but the law itself was not enough. Each relevant agency needed to write regulations to implement the law, and the amazing thing is that this process of adopting regulations remains incomplete in 2020.

According to Julia Brunner of the National Park Service (NPS) one major cause of the delay is the number of agencies involved in different government bureaus; principally these are the Bureau of Land Management (BLM) and the National Park Service; parts of the Department of the Interior (DOI); and the National Forest Service, a unit of the Department of Agriculture.[32] Although the agencies collaborated in writing the regulations, each had its own legal mandates and rules that initially were not known to the other agencies. In addition, each set of rules was subject to public comment, including from academic paleontologists, amateur collectors, and commercial collectors. The result was a set of similar, but not identical regulations. The Forest Service rules were the first put into effect, in April 2015.[33] The DOI agencies rules were published for comment in late 2016 and are still awaiting final approval.[34] As

Brunner told me, the new regulations are transparent, consistent among the agencies, and more organized and efficient than existing practices, making obtaining permits easier. She added that they should allow better collaboration between the agencies and researchers. These regulations have real teeth in terms of criminal penalties and are all laudable goals that should lead to much better protection of paleontological resources, especially vertebrate fossils, on public lands.

The Forest Service regulations cover fifty-two pages of three-column text in small type; many pages contain responses to the numerous public comments (177!) on the act as a whole or on specific sections. One key aspect of the regulations is the distinction between research, which requires a permit, and "casual collecting," which is defined as "the collecting without a permit of a reasonable amount of common invertebrate or plant paleontological resources for non-commercial personal use, either by surface collection or the use of nonpowered hand tools, resulting in only negligible disturbance to the Earth's surface and other resources." A reasonable amount is described as twenty-five pounds per day or one hundred pounds per year, including matrix, and personal use excludes the use in research or for sale. Vertebrate fossils of any kind, such as shark teeth, are prohibited from casual collecting. "Non-commercial personal use" includes not only sale but also research. And by "casual" they mean "an activity that generally occurs by chance without planning or preparation." For educational purposes, students collecting for themselves on a field trip would be considered casual, whereas a specimen collected by the instructor for a teaching collection would not and would require a permit.

It is not surprising that the concept of casual collecting received the most comments. Concerns include that the rules are too restrictive, that "common" and "casual" are poorly defined, and that there are insufficient personnel to enforce the rules. My own reaction, as somebody who often works with very heavy slabs of limestone, is that defining a "catch limit" based on weight is impractical. I also quibble with the idea of "casual": I go to a collecting site because I already know it is there, which requires planning and preparation. I assume that the same is true for any amateur, student, or professional. And I am not happy that a common invertebrate specimen I find and decide to put in our teaching collection would trigger the necessity of a permit.

There was also a great deal of comment about the requirement of "confidentiality of specific location information for paleontological resources," and "does not distinguish among vertebrate, invertebrate, plant, common, abundant, uncommon, and/or rare paleontological resources" with release from confidentiality up to an agreement with an agency employee, the authorized officer. Confidentiality is understandable for many vertebrate sites because publicly available information could lead to poaching. At the same time, detailed published locality information is critical for many paleontological studies. In this case, I trust the scientists themselves, in their own interest, to protect the confidentiality of their sites if they believe it is necessary.

And, of course, there is the paperwork, which can be very time-consuming. Lindsay Zanno, a vertebrate paleontologist at the North Carolina Museum of Natural Sciences, tells me that her lab staff spends four to six weeks each year dealing with required permits and reports.

In a presentation given at the 2015 meeting of the Geological Society of America, ten academic invertebrate paleontologists objected strongly to the published Forest Service regulations, suggesting that they are too heavily influenced by the concerns of vertebrate paleontologists: "Academic invertebrate paleontologists routinely depend on avocational collectors and generally do not object to commercial trade. No information comes from an undiscovered fossil."[35] They add that the collection and sale of common fossils often uncovers rare ones, providing access to these finds for academic paleontologists who may lack funding (as is the case with Lance Grande). They conclude that "the new regulations greatly increase costs and restrict the kinds of research that invertebrate paleontologists can conduct on public lands. They also result in legal obscenities, like allowing large quantities of fossil-bearing rock to be processed into concrete or other stone products, while prosecuting a collector for taking a few hundred specimens from the same rocks."

Amateur paleontologists also expressed a great deal of unhappiness. Linda McCall is the winner of the 2020 Strimple Award of the Paleontological Society, which is given to an amateur who has significantly furthered the field and has authored papers in professional journals.[36] As president of the North Carolina Fossil Club, she sent a comment on the proposed regulations, pointing out that under the proposed rules "our entire segment of the paleontological community is effectively disenfranchised."[37] Specific objections address the definition of casual collecting to include research, because amateurs would then feel precluded from collecting something unusual and thus furthering science; the use of "common," because this is unworkable; and the weight limits as being too restrictive and making it difficult to collect enough fossils for outreach events. Reasonable replacement language is proposed. Linda has told me that "NONE of the papers I lead authored could have been written if the specimens had been on public land and PRPA had been in effect in its current form."

Personally, I object to the idea that every fossil is scientifically valuable. Many fossils are indeed unique and precious and require protection from exploitation. At the same time, the vast majority of fossils are incredibly common, and many buildings are made using fossil-bearing limestones. Near where I live, quarries remove and crush tons of fossiliferous rock every day for gravel and cement (figure 16.1). Fossil collecting is a wonderful activity for

FIGURE 16.1 Fossiliferous dolomite being crushed at Thornton Quarry, Illinois.
Source: Photo by the author.

children and is a breeding ground for future generations of paleontologists. Access to fossil collecting sites is essential for teaching paleontology. And I am in agreement with the Paleontological Society Code of Fossil Collecting: "To leave fossils uncollected assures their degradation and ultimate loss to the scientific and educational world through natural processes of weathering and erosion." We need to encourage knowledgeable collecting and good relationships with amateur and commercial collectors to increase the number of fossils collected and to make sure scientifically valuable ones end up in the right hands.

Native American lands are expressly omitted from the PRPA, with control of their fossil resources being left to the tribes. This is a welcome change given the long history of removal of fossils from native lands without permission or compensation.[38] This is part and parcel of the sad record of resource exploitation and removal of archaeological remains, which is finally receiving the public attention it deserves, such as in the PBS series *Prehistoric Road Trip*.[39] In *Dinosaurs and Indians*, Lawrence Bradley focuses on the history of dispossession of paleontological resources from Sioux lands, which contain some of the most important fossil sites in the world.[40] This history includes the important figures of nineteenth-century vertebrate paleontology in the United States, such as Joseph Leidy, J. B. Hatcher, Edward Drinker Cope, and Othniel C. Marsh. With some exceptions, the removal of fossils without permission continued well into the twentieth century. As Allison Dussias puts it, "the history of encounters between Native Americans and paleontologists and other scientists . . . indicates a systematic disregard for tribal sovereignty and treaty rights."[41] Bradley tracked down many of the fossils from Sioux lands that are now in major museums; virtually none record that they were taken from a tribal territory. The sale of SUE also remains an issue, with the tribe receiving nothing from the sale. In an article that discusses the legal issues of the case, Allison Dussias points out that the court's opinion in the case "did not discuss Native American attitudes toward the land or toward stones and fossils in determining rights in the fossil." Critically, the ruling also established that the federal government holds paleontological resources found on tribal lands in trust for Indians and the tribe. Bradley insists that "tribes need to manage their own paleontological resources," and Dusias lays out the legal basis to do so. Steps are now being taken in this direction. The Standing Rock Sioux tribe in 2007 published a Paleontology Resource Code (emended 2015)[42] that established a Paleontology Department and the Standing Rock Sioux Institute for Natural History (closed in 2019 because of a lack of funding). The code is blunt; paleontological resources found on tribal lands belong to the tribe. It also lays out the responsibilities of the Bureau of Indian Affairs vis-à-vis fossils on lands it holds in trust for the tribe.

Of course, it is incumbent on paleontologists to recognize this history of appropriation of paleontological resources, not only from Native American lands but from other countries as well, to repair our relationships with indigenous peoples. This may not require repatriation, but it does require recognition of the tainted origin of many significant fossils. It mandates, as laid out in the Paleontological Society Code of Fossil Collecting, that we collect in compliance with tribal laws. We must view indigenous communities as partners in our research and recognize, through active engagement, their voices on the land and on the fossils. And it creates a moral necessity that we encourage the development and training of Native American paleontologists, who are markedly unrepresented, and participation by paleontologists in meetings of the Society for Advancement of Chicanos/Hispanics & Native Americans in Science (SACNAS). The Paleontological Society has been a visible presence at SACNAS for several years; I am pleased that in 2020 it announced grants to support attendance at that meeting by members from underrepresented or at-risk groups.

One area where tribal and paleontological interests coincide is the fate of the Bears Ears and Grand Staircase–Escalante National Monuments, both in Utah. Grand Staircase–Escalante National Monument was established on nearly three thousand square miles of federal land in southern Utah in 1996 by order of President Clinton, over strenuous local political and public opposition (figure 16.2). The area has immense hydrocarbon reserves, and there was a fear of destructive mining activities. The proclamation creating the monument noted that it contained "world class paleontological sites," as well as significant archeological sites and usage by modern indigenous cultures.[43] The promise of the monument for paleontology has been realized. As pointed out by David Polly, it contains one of the best records we have of Cretaceous ecosystems up to the extinction of the dinosaurs. He estimates that nearly half of SVP members use data gathered there.[44]

The more than two thousand square mile Bears Ears Monument was created in 2015 by President Obama, after years of lobbying by Hopi, Ute Indian,

FIGURE 16.2 Grand Staircase—Escalante National Monument, Utah.
Source: Photo by the author.

Ute Mountain Ute, Zuni tribes and the Navajo Nation, with the chief goal of protecting religious and archeological sites.[45] It is important to note that the tribal governments were directly involved in management of the monument. Obama's proclamation also recognized that "the paleontological resources in the Bears Ears area are among the richest and most significant in the United States." Research in the area has indicated that it contains critical sites for understanding the Permian and Triassic extinctions, including a huge concentration of Triassic fossils.

Unfortunately, the promise for science at both monuments is now threatened. Less than a year after its creation, President Trump reduced the size of Bears Ears by 85 percent and reduced the size of the Grand Staircase—Escalante by nearly half. The tribes immediately sued to block the reduction at Bears Ears,[46] and they were soon joined by conservation groups and the SVP. The SVP and local conservation groups are also suing at Grand

Staircase—Escalante. The situation is still in flux. In January 2021, President Biden ordered a sixty-day review of the monuments and their boundaries; it is likely that they will be restored.[47]

Not all public lands are owned by the federal government; many state and regional parks also contain fossils. Some of these allow collecting, others do not. Collecting at Illinois state-owned sites of the Mazon Creek biota has been hampered by the unfamiliarity of land managers with fossil collecting compared to fishing and camping. Along with colleagues, I am working to convert a former quarry to a state park administered "fossil park," open to collecting by educational and hobbyist groups. Similar parks exist elsewhere, such as in Texas, Iowa, and New York.[48]

One important aspect of collecting on public lands and occasionally on private lands is mitigation paleontology. As is the case with archeological remains, fossils are often found during the process of construction; in the Sioux tribal code these are "inadvertent discoveries." Sometimes, as is the case with the Mammoth Site of Hot Springs, South Dakota, and the St. George Dinosaur Discovery Site at Johnson Farm, Utah, construction is abandoned, and the site becomes a museum and excavation site. Other times, however, fossils must be removed ahead of the bulldozers. In 2010, construction of a reservoir at Snowmass Village, Colorado, uncovered a huge trove of ice age fossils. Working with astounding speed, a group led by Kirk Johnson uncovered tens of thousands of bones belonging to some fifty vertebrate species, as well as insects and plants, before the site was covered by the waters of the reservoir.[49]

In many states and municipalities, uncovering fossil remains requires that construction be halted and that a trained and certified "mitigation paleontologist" be brought in to assess the find, salvage them if necessary, and making sure they are properly stored and curated.[50] In California, which has particularly strong laws for mitigation paleontology, warehouses are filled with fossils uncovered during construction. In their terrific book *Cruisin' the Fossil Coastline*, paleobotanist Kirk Johnson and paleoartist Ray Troll describe visiting the Cooper Center in Orange County: "The active construction sites in the area have, since the 1970s, produced countless fossils carefully monitored by county-mandated professional paleontologists. The many fossils they collected ended up in this odd facility [that] was all collection and no museum."[51]

Issues involving collecting do not just involve vertebrates or public lands. Access to private lands is influenced by the "SUE effect" and by concerns over liability, and many landowners are now hesitant to grant access. In my own work, I have found it increasingly difficult to find quarries willing to give me entry for research or teaching. Much of this is due to fears of legal liability or enforcement action from the U.S. Mine Safety and Health Administration. Economic conditions have also closed many quarries, or they have been converted to other uses.

Fossils are wonderful objects. They represent the best evidence for the long and complex history of life on Earth. They are invaluable tools in our ongoing search for fossil fuels and in our efforts to reconstruct the geological and environmental history of the Earth. They excite the imaginations of children and adults alike. They also are, in a few noteworthy cases, worth substantial sums of money. Paleontologists suffer from the fact that the objects of our scientific interest are also objects of pecuniary interest.

When I was an undergraduate, I had the opportunity to be a work-study student at the American Museum of Natural History, working with Niles Eldredge. An integral part of a museum scientist's job, I soon learned, was interacting with members of the public, in particular with individuals who had fossils they wanted identified. They all had three questions: "What is it?," "How old is it?," and inevitably, "How much is it worth?" Although we are glad to answer the first two questions, we would not and should not answer the last one. To a paleontologist, the only true worth of a fossil is its scientific value.

17

FOR THE LOVE OF FOSSILS

T he state fossil of Illinois is the bizarre Mazon Creek animal *Tullimonstrum* (the Tully monster). Although the name sounds like something out of Sesame Street, it was named after local fossil collector Francis Tully, who in 1958 showed his odd fossil to Eugene Richardson and Ralph Johnson at the Field Museum. Ten years later they described it, largely based on the "thousands of specimens collected by amateur and professional paleontologists," with the genus name formalizing "the usage of the amateur collectors, who called it the 'Tully Monster.'"[1]

Tully was one of the legions of avocational paleontologists, such as George Langford Sr. (figure 17.1), who have collected fossils from the Mazon Creek area over the last century.[2] Tens of thousands of specimens have been donated to museums, and these have formed the basis for many of the hundreds of professional papers describing the fossils and environment of Mazon Creek. Much of this positive interaction was facilitated by the warmth and openness of Gene Richardson; as described by John Catalani: "Gene took the time to talk with, not at, the amateur collectors, making them feel that their efforts in collecting Mazon Creek sites and their subsequent donation of specimens was an important contribution to the science of paleontology."[3] The

FIGURE 17.1 Long time Mazon Creek collector George Langford Sr. in 1938.

Source: Photo courtesy of George Langford III.

definitive guides to the plants and animals of the area were written by amateur paleontologists and published by the Earth Science Club of Northern Illinois (ESCONI),[4] whose members include the acknowledged experts on Mazon Creek. ESCONI has been honored by having several Mazon Creek fossils named after it, such as the enigmatic fish *Esconichthys*. Their knowledge goes beyond Mazon Creek; they are also familiar with fossils and fossil sites throughout the area. I am an enthusiastic member and rely on their knowledge and access to sites.

The members of ESCONI are exemplars of those who do not make their living from paleontology but for whom the study of fossils is their passion and their hobby. Apart from amateur astronomers and birders, no group of avocational scientists have contributed as much to science as amateur paleontologists. Many spend every spare moment collecting and eagerly share their best finds with the professionals, and they donate them to museums. Some write papers, frequently with professionals, that are printed in professional journals. For example, one of the most prolific authors of papers on eurypterids (sea scorpions; the subject of my dissertation) and fossil scorpions was Erik Kjellesvig-Waering, by profession a petroleum geologist.[5] Many of his papers were coauthored by the professional paleontologist Kenneth Caster of the University of Cincinnati. Another enthusiastic eurypterid and scorpion collector is retired chemist Samuel Ciurca of Rochester, New York. A member of the Buffalo Geological Society and the Fossil Section of the Rochester Academy of Science, the bulk of his immense collection, by far the largest in the world, was donated in 2006 to the Yale Peabody Museum.[6] Similarly, Linda McCall of North Carolina has collaborated with professionals, given presentations at national meetings, and has donated huge quantities of fossils to museums.[7]

Another example of an invaluable amateur is Mohamed 'Ou Saïd' Ben Moula of Morocco. In the early 2000s, he discovered an Ordovician fossil locality known as the Fezouata biota. This locality is remarkable for the preservation of soft-bodied organisms and is a key window into the evolution of life in this critical interval of Earth history. It has been a subject of intense interest. Not only did Ben Moula bring the attention of professionals to this site, but as stated in the introduction to an entire volume on it: "without him, and his eagerness to support scientific research, this volume would not exist, and no-one would ever have known about this major exceptionally preserved fauna.... Not only did Mohamed discover the Fezouata Biota and recognized its interest—crucially, he has also always been very open and supportive of scientific research on any of the fossils discovered by him. Last but not least, he is an absolutely remarkable collector, recovering mm-sized fossils that are almost invisible to the naked eye."[8]

ESCONI is one of sixty local fossil clubs in the United States, and it has just celebrated its seventieth year.[9] The oldest of these organizations is the wonderfully named Dry Dredgers of the Cincinnati area, established in 1942. The Cincinnati area is renowned for numerous well-exposed and highly fossiliferous outcrops of Ordovician age rocks.[10] Serious fossil collecting there

began in the mid-nineteenth century and continues to this day. The early collectors, known informally as the Cincinnati school, not only amassed huge
numbers of fossils but published numerous papers on them.[11] Some, such as
Charles Schuchert, developed into professionals. Schuchert became professor
of paleontology at Yale and is the academic ancestor of a large proportion of
American invertebrate paleontologists, me included.[12] Members of the Dry
Dredgers have continued this tradition of publication into the present day,
working closely with professionals, especially with those in southern Ohio
and northern Kentucky. Like ESCONI, specimens collected by members of
the Dry Dredgers are often donated to regional natural history museums.

One of the largest clubs, with five hundred members, is the Denver area
Western Interior Paleontological Society (WIPS), founded in 1985. As is the
case with ESCONI and the Dry Dredgers, they sponsor field trips and publish
a detailed monthly newsletter. Every other year since 1998 they have sponsored a "Founders Symposium" focused on topics of paleontological interest,
most of the speakers being professional paleontologists. Topics have included
the "Fossils & Flight," "Extinction," and "Exploring the Morrison Formation."
In addition to the speakers, there are also exhibits and displays of paleoart.

Until recently, these local avocational groups worked in isolation; there
was no national umbrella society, such as the Astronomical League for amateur astronomers. There were also no clear-cut pathways for amateur clubs to
communicate with professional paleontologists or to make the knowledge of
both groups readily available for outreach to schools. In response to this, in
2013 the Florida Museum of Natural History began The FOSSIL Project,[13] with
the "aims to enhance communication between fossil club members and professional paleontologists, engage club members in training and development,
allow club members to attend meetings and workshops, and conduct K–12
outreach to underserved audiences." Also calling themselves "social paleontology," the project took a multipronged approach to building a "community of practice" comprising both professional and amateur paleontologists.
Central to this was creation of the myFOSSIL website, with numerous links,
including to a site for members of the community to post digital images of
fossils they had collected as well as 3-D scans. The project had a highly visible social media presence and highly informative newsletters that highlighted
clubs and individual amateurs and professionals. There was also a strong
effort at K–12 outreach. Unfortunately, support from the NSF grant ended in

September 2019, which meant they could no longer support a full-time paid staff. They are hoping to maintain the momentum using volunteer members of the community. I hope they succeed.

Avocational paleontologists have long been leaders in paying it forward to the profession and to the wider community. Many volunteer at natural history museums, where they are invaluable contributors. Some act as docents, others work in preparation labs or assist with cataloging collections. WIPS members are helping catalog the collection of the Geology Museum of the Colorado School of Mines. Paleoartist Wendy Luck is a docent at the Denver Museum of Natural Science's Prehistoric Journey exhibit and a volunteer "Leaf Whacker," prepping paleobotany fossils in their collections.

Other amateurs are active in outreach, encouraging future generations of scientists. For example, Bruce and René Lauer run the Lauer Foundation for Paleontology, Science, and Education (PSE)[14] and do numerous presentations in area schools and at STEM programs that heavily utilize the foundation's amazing fossil collection. In 2019, they interacted with an estimated three thousand K–12 students. Recently, they purchased a Triassic microvertebrate site in New Mexico and are making it available for students and faculty from Appalachian State University and the Virginia Polytechnic Institute, and researchers from the Natural History Museum in London.[15]

Amateur societies also provide funding in various forms. The Mid-American Paleontology Society (MAPS) annually supports the Paleontological Society Student Research Grants, where the top three proposals receive MAPS Outstanding Research Awards. ESCONI also supports Paleontological Society student grants. The Western Interior Paleontological Society is a donor to *Palaeontologia Electronica*, an online professional paleontology journal. Both WIPS and the Dry Dredgers have their own small grant programs.

One of the ways fossil clubs support their activities is through fossil and rock shows of various kinds. ESCONI has an annual Gem, Mineral, and Fossil Show that features exhibits, live auctions, and outside vendor sales. One of the largest shows is the National Fossil Expo, run by MAPS, and it is for fossils only. Again, there are items both for display and for sale, and presentations by both amateurs and professionals. An auction benefits the Paleontological Society grants.

Like Gene Richardson, most professionals embrace and honor members of the amateur community. Since 1984, the Paleontological Society (PS) has

awarded the Harrell L. Strimple Award for "outstanding achievement in paleontology by amateurs (someone who does not make a living full-time from
paleontology)."[16] Strimple (1912–1983) was a self-trained paleontologist who
became the longtime curator of the paleontological collection at the University of Iowa. He was incredibly prolific, with more than three hundred scientific publications. He was also an important contributor to the *Treatise of
Invertebrate Paleontology*, the "encyclopedia" of nonvertebrate animals. Similarly, the Paleontological Research Institution (PRI) began in 1993 to present
the Katherine Palmer Award to recognize "an individual who is not a professional paleontologist for the excellence of their contributions to the field.
This award is named for PRI's second director, Katherine Palmer, who held
avocational paleontologists in high regard and collaborated with many during
her long career." This award is presented at the Mid-America Paleontological Society meetings. The Society of Vertebrate Paleontology (SVP) gives the
Morris F. Skinner Award to honor those who make "important collections of
fossil vertebrates" as well as "persons who encourage, train or teach others
toward the same pursuits."

Many members of the SVP are amateurs, as are attendees at their annual
meetings. In 2017, the Paleontological Society established the Avocational
Paleontologist Liaison to the PS Council, with Jayson Kowinsky (aka Fossil-
Guy) being the first named to this position. By profession, he is a high school
physics teacher; by avocation he is an avid fossil collector and a champion
of outreach, including a truly useful website.[17] The PS is also now actively
encouraging amateurs to join.

Some avocational paleontologists have transitioned to making their livelihood using their enthusiasm and knowledge. Riley Black, who previously
wrote under the name Brian Switek, was the longtime author of the popular
Laelaps blog for *Scientific American*. She has "been blogging for over a decade
for outlets such as ScienceBlogs, WIRED, the Smithsonian, *National Geographic*, *Scientific American*, and Patreon, with bylines in publications ranging
from *Slate* to *Nature* and *Nerdist*. I've also written several books about fossils,
and these tomes have led me to speaking gigs for audiences across the country. I've written scripts, museum blog posts, tweets, and so much else, trying
to use every tool I come across to share paleontological discovery. . . . Even
though I'm not a professional paleontologist, I appreciate that full-time paleos
have encouraged people like me to become involved and contribute."[18]

Joseph J. "PaleoJoe" Kchodl, 2001 winner of the Katherine Palmer award, has had a long love affair with trilobites and has developed a traveling exhibit on them. PaleoJoe also presents numerous programs at schools, clubs, and senior centers and says: "No one is immune from learning and enjoying pale-ontology!" Kchodl also leads fossil collecting trips in Michigan and elsewhere and sells high-quality fossil specimens online.

Of course, not every amateur works well with professionals. Some are reluctant to donate a rare find. And many lack the training to gather and record necessary collection data.

A final issue is what to call those who collect fossils as a hobby. At least one of my colleagues objects to the term "amateur" because of its negative connotations of incompetence. Suggested alternatives include "avocational," "nonprofessional," "hobbyist," or "citizen." As a longtime amateur astronomer, I do not object to the adjective. And I also recall the long-running column "The Amateur Scientist" in *Scientific American*. But it is the members of a group who should choose their label. ESCONI member Jack Wittry (2015 Strimple Award winner) prefers "amateur," and fellow member John Catalani (2000 Katherine Palmer award winner) also favors amateur and "will accept avocational, but I reject non-professional or rockhound." John was the long-time writer of "An Amateur's Perspective" in the *American Paleontologist*, a now defunct publication of the Paleontological Research Institute. In a commentary in the online journal *Palaeontologia Electronica*, John wrote that he favors "amateur because it indicates why we started collecting—for the love of the activity and without remuneration." In a survey reported as part of The FOSSIL Project, 57 percent of respondents preferred amateur. Linda McCall is the current Advocational Liaison to the Paleontological Society and prefers that term, but she has no problem with amateur. So let's use both, respecting the individual's preference.

18

PRESENTING THE PAST

My wife and I were at the visitor's center at a state park, and at the urging of his parents, a young boy (about seven years old) timidly asked me, "Are you a paleontologist?" When I said that I was, he loudly squealed in joy. What gave me away was that our car had a bumper sticker from the Paleonto- logical Society that says "Teach the Science!" The youngster then told me that when he grows up he wants to be a paleontologist. I assured him that I thought it was a wonderful idea. And then he asked me, "How many dinosaurs have you discovered?" Oops. Not being one to lie to children, I had to tell him the truth: that I do not study dinosaurs, but I looked at some cool fossils that were not dinosaurs. That was clearly not what he wanted to hear; the bloom was off the rose. He walked away. (By the way, even though his mom was a scientist and was wearing a March for Sci- ence T-shirt, they still assumed that I was the paleontologist, not my wife. Maybe it's the beard; see figure 18.1).

There is no doubt that our biggest fans are children, especially young children. Children's toys and books about dinosaurs are ubiquitous. Media also caters to this interest. There have been numerous dinosaur-themed movies, including *Land Before Time;* *Ice Age: Dawn of the Dinosaurs;* and *The Good Dinosaur. Dinosaur*

FIGURE 18.1 The archetypical bearded paleontologist in the field, not looking for dinosaurs.

Source: Photo courtesy of the author.

Train is a very popular program on PBS. I have also gone down this road; years ago a theater group my wife worked for asked me to join the cast of a touring production to schools called *Do You Really Want a Dinosaur?* The premise was that a young girl wanted a pet dinosaur, but her mother told her that she first needed to learn all about dinosaurs. That was where I came in; I played a museum paleontologist and gave a slide show about dinosaurs and answered questions from the children in the audience. My one acting gig; my role was not much of a stretch.[1]

Jennifer Bauer (University of Michigan Museum of Paleontology) describes the public interest in paleontology this way: "The exposure to paleontology happens early and really excites young minds—something I think is unique to paleontology due to the exposure in popular media and [the] grandeur of dinosaurs. Paleontology is something that all age groups and demographics either have a personal connection to or can generally get excited about. This makes it very easy to connect and share information with folks ages 2–65+. It is also a very tangible subject; you can hold fossils and rock specimens and get a sense of what it once was. Personally, I have a hard time rationalizing producing research if I don't spend time making sure folks can understand the importance of doing it."

Similarly, Herb Meyer (National Park Service; box 18.1) says, "Paleontology has become very popular with people, especially kids, even more so than when I was young. It's fascinating to meet the kids who want to become paleontologists, and they seem so intrigued to meet one. We have programs that offer three different kinds of 'badges' for kids based on completing exercises in paleontology booklets, including 'Junior Paleontologist' badges. We offer a number of youth-centered education programs to help people learn at an early age."

BOX 18.1

Herb Meyer is a paleobotanist working for the National Park Service at Florissant Fossil Beds National Monument. He first became interested in geology while in junior high school in Oregon and was a member of a local geology club for kids; "we took many field trips to fossil collecting sites. Being able to collect fossils at an early age is a lot of what inspired me for the career. I developed an interest in fossil plants very early, and I think part of the fascination came from splitting pieces of rock to find leaves of plants that were beautiful, intriguing, and entirely different from the forest in which I was sitting while pounding on rocks." While still in high school, Herb spent three summers at a geology-oriented summer camp, which also gave him many opportunities for fossil collecting. A high school research project ended with him being one of forty finalists for the Westinghouse Science Talent Search, which led to a trip to Washington, D.C. and a chance to meet with paleontologists at the Smithsonian. He attended Portland State University but eventually transferred to the University of California, Berkeley and completed his undergraduate

degree there. "I already knew a lot of the people there and had vis-
ited when I was in high school, and Berkeley always seemed like the
paleontologists' mecca to me."

Herb eventually got both his master's degree and PhD at Berke-
ley. "I'm still amazed today at how many of my contemporaries as
grad students I see on television documentaries about paleontology,
and what a great opportunity we had to share at Berkeley. There's
no doubt that being in such a dedicated paleo program was abso-
lutely inspirational in my career." Herb himself has mentored many
younger scientists through internships in the national park, such as
Gwen Antell, who is completing her PhD at Oxford. His job as a
paleontologist at the National Monument involves "researching and
conserving late Eocene (34-million-year-old) plant and insect fos-
sils. . . . I study climate change based on fossil leaves, which helps
to answer other questions such as how high the elevation was in
the Eocene, how ancient organisms responded to these changes, and
how modern climate change might affect biotic communities today."

As discussed in chapter 15, one key component of a National Science Foundation (NSF) proposal is "Broader Impacts"; that is, how does your proposed research benefit society in some way? For many of us, this is expressed through the mechanisms of education and outreach. The former covers all areas of formal education, from kindergarten to graduate school, as well as informal education, such as that conveyed through a museum exhibit. These may be focused on our discipline, or more broadly in science, technology, engineering, and math (STEM). A welcome recent emphasis is on mechanisms for increasing diversity in our markedly not diverse science. I have been a member of several NSF funded projects that address this issue, involving close interactions with teachers and students from Chicago Public Schools, such as developing an earth science course for high school students that gives them college credit. I have also helped revise how we teach elementary science to preservice teachers, especially in the context of the recently developed Next Generation Science Standards.[2]

Outreach is broader, encompassing not only education but all the ways we communicate our science to the broader public. Community engagement can include giving lectures, developing informative websites, blogging, answering questions from the public, holding a fossil identification day at a museum, or writing books like this one.[3] I earlier discussed the outreach activities of many museums.

Some of my colleagues have made education and outreach a cornerstone of their careers while still pursuing cutting edge research programs. Rowan Lockwood (College of William & Mary; box 18.2) is currently education and outreach committee chair for the Paleontological Society (PS). One constant throughout Rowan's undergraduate and graduate career was an interest in education and outreach. While at Yale, she started a community outreach organization to bring a hands-on introduction to science to inner city middle school students in New Haven. This was later expanded to the University of Chicago, Loyola University, and North Park University in Chicago. She also volunteered at the Education Department at the Field Museum. Rowan is a renowned educator, being named one of the top three hundred professors in the country by the *Princeton Review*. In her time at the PS, Rowan has championed initiatives to increase diversity in the science and in the society's leadership. She has also spearheaded efforts to improve the teaching of paleontology. All this while doing important research in conservation paleobiology.

BOX 18.2

Rowan Lockwood grew up in Rockford, Illinois. She hated science in high school: "too much rote memorization and cookie cutter lab experiments!" However, to graduate she had to have an independent science project. History was her favorite subject, so she chose the history of life. Rowan read about research that had been done on pterosaurs and found it so fascinating that she reached out to two vertebrate paleontologists to advise her (Virginia Naples at the nearby Northern Illinois University and Kevin Padian at Berkeley). Her project literally and figuratively took off, and she "developed a model to calculate how much femoral muscle mass would be required for a pterosaur like Pteranodon to take off from a bipedal position on the ground." This project led to a top forty Westinghouse Science Talent Search award, a talk at a Society of Vertebrate

Paleontology conference as a senior in high school, and a Smithsonian internship at the Smithsonian. "[I was] really surprised how much I enjoyed research and the scientific community. So different from what you learn about in textbooks or see on TV." Rowan went to Yale, originally majoring in history but eventually switching to a double major in geology and organismal biology. This led to a master's degree at Bristol (in the UK) and a PhD at the University of Chicago. She states, "[I] would not have had the confidence to finish grad school without the help of my peer students." Rowan's love for teaching led her to look for jobs at liberal arts colleges rather than at a research institution, and she has focused her efforts on undergraduate research and education.

Another leader of the workshop and short course was Margaret Yacobucci (Bowling Green), a champion of STEM education. Peg has campaigned to bring modern pedagogical methods into the paleontology classroom, notably understanding how preconceptions impact student's learning. Peg was the PS education and outreach coordinator from 2010 to 2013. She is also a world acknowledged leader on the paleobiology of ammonoids, a group related to modern octopus, squids, and *Nautilus*.

The efforts of Rowan and Peg and others have led to an ongoing shift in how college-level paleontology is taught. We can roughly divide university instruction in paleontology into three eras. The first era, lasting most of the twentieth century, was characterized by training students to be able to identify as many fossils as possible, often to the genus level. This was driven by the dominant use of fossils for geological purposes, such as dating rocks (biostratigraphy) and determining environments (paleoecology). A typical textbook of this era, such as *Invertebrate Fossils*,[4] has a short introductory section on principles and methods, with the remainder of the book providing a group-by-group review of fossils. There is a lot of morphologic and taxonomic terminology and only a few pages on evolution. Older geologists remember having to slog through such courses, leading to the misconception that memorizing fossils is what paleontologists do.

The first sign of a new era dawning was the 1960 publication of *Search for the Past: An Introduction to Paleontology* by James Beerbower.[5] The first half of the book was dominated by what we would now call paleobiology, and the second half provided a mercifully simplified review of vertebrates and invertebrates. In a review of the first edition, David Raup identified its main strength as representing paleontology as a dynamic field, offering great opportunities for new research directions. I remember it as being lively and fun.

The flood burst in 1971 with Raup and Steve Stanley's *Principles of Paleontology*.[6] Except for an example of how it is done, there was not a fossil description to be seen. It was a book of theory, methods, and concepts. Raup and Stanley introduced a whole new way of looking at paleontology: the focus was not on the fossils themselves but on fossils as critical data for looking at big questions in the history of life. They assumed you could look at the older texts to learn the fossils. The textbook was a key part of the paleobiological revolution then sweeping through paleontology, in which Stanley and Raup were key players (see chapter 7). In a true Kuhnian sense, it and its later editions solidified the new paradigm.

The most popular undergraduate textbooks currently out are descendants of both Beerbower and Raup and Stanley. Both Don Prothero's *Bringing Fossils to Life* and Michael Benton and Dave Harper's *Introduction to Paleobiology and the Fossil Record* are about equally divided between concepts and descriptions of the fossil groups. Both contain a heavy dose of modern paleobiological theory. This approach is mirrored in my own teaching; students need to know what a brachiopod is before they are introduced to the concept of brachiopod clade dynamics over time, but they should know both.

The third era of paleontological teaching, which is dawning now, is different not for what is taught but for how it is taught. The revolution in earth science pedagogy is driven by concepts coming from the education and learning science communities, and it is led by organizations such as the National Association of Geoscience Teachers and the Paleontological Research Institution and individuals such as Yacobucci and Lockwood. We now have a far more nuanced understanding of how people learn. The shift can be seen by looking at the programs for a 2009 workshop on *Teaching Paleontology in the 21st Century* and a highly successful 2018 PS short course on *Pedagogy and Technology in the Modern Paleontology Classroom*.[7] The workshop focused on the best strategies for teaching paleontology, including specific examples, as well as how paleontology can be incorporated into the general science curriculum. The short course included presentations on topics such as teaching with online databases,[8] "flipping" the classroom,[9] applying educational research to teaching and addressing prior conceptions,[10] and dealing with issues of equity, culture, and the "sense of place" of students.[11] Current and future generations of instructors will teach paleontology far differently than I did, and I believe they will teach it far more effectively!

Adoption of the Next Generation Science Standards in many school districts has resulted in a new emphasis on evolution and earth science.[12] We have never properly capitalized on children's interest in dinosaurs and other prehistoric life. I am not talking about money, although I have often wondered how much dinosaur-themed merchandise and media earn compared to the amount spent on dinosaur and other fossil research. What I am interested in is the potential of fossils as a hook to promote a lifelong interest in science. Paleontology, an integrative and multidisciplinary science, can be used as a focus for teaching many other sciences.

We can use fossils to teach about the process of science. What really impresses me about kids is how much they know about dinosaurs. They know the long Latin names. They know what dinosaurs ate and how we know that. We can use this knowledge as a springboard to explore the way science asks and attempts to answer questions about nature. We can use fossils to teach biology. Basic questions about physiology, anatomy, ecology, and evolution can be taught using ancient life as an example. We can use fossils to teach chemistry. The composition of bones, teeth, shells, and the rocks that contain the fossils can be used to introduce basic concepts of stoichiometry. Weathering and taphonomy of the remains are examples of reactions. We can use fossils to teach physics. The biomechanics of jaws are examples of machines. Locomotion on land or water can be used to discuss the mechanics of motion. We can use fossils to teach environmental science. Changes in the distribution of animals over time or in the shapes of leaf margins illustrate how life responds to climate shifts. These are just some examples that come immediately to mind. Instead of teaching biology, physics, chemistry, geology, and environmental science as separate fields, each in its own silo, why not emphasize how they blend and interact with each other? Why not put fossils (okay, dinosaurs) at the center and show how each discipline of science illuminates one aspect of our understanding?

The internet has produced a sea change in both education and outreach. Sarah Sheffield is an assistant professor at the University of South Florida with a research focus on early echinoderms. Like many other paleontologists, she came to the field with an initial interest in the arts, in her case, music: "I went to UNC Chapel Hill to be a music performance and music education major. Almost immediately, I realized that I didn't have the passion that was necessary to be a musician. . . . I loved music as a hobby, but the thought of doing it every single day made me lose that passion I once had. I took a course (Prehistoric Life) that was a biology and geology listed course and that was it for me. After the first day, I fell in love with the material and the storytelling that came with teaching the story about life's evolution. I was offered a chance later to teach a lab section of this class as a junior, and that's really when I decided I had no other career path but to be a paleontology professor because I just loved teaching so much." Sarah loves working with kids but believes it is critical, "if you have the opportunity to explain why science is important to someone who votes, then that's a great time to explain to them why climate change is a priority or why paleontology has real meaning in this world." Sarah uses various approaches to reach the public, for example, using the "program *Skype a Scientist* a lot to talk to K–12 students. This is a great way to get involved—I've video chatted with second graders up to students coming back to school in their twenties to earn their GED—all of them have been great ways for them to ask questions and for me to tell them a bit about what I do." She also blogs for the site *Time Scavengers*.[13]

Time Scavengers is a wonderful site, established by Jen Bauer and Adriane Lam, with the avowed aim of "providing the general public with a basic understanding of our most important geological concepts, including those that are important for understanding climate change and evolution."[14] Jen is the research museum collection manager at the University of Michigan Museum of Paleontology and is a specialist in Paleozoic echinoderms. Adriane is a postdoc at Binghamton University and is reconstructing Cenozoic oceans using foraminifera. The site is a treasure trove of accessible information on geology, paleontology, and climate change. Jen describes it this way: "We aim to reach everyone—normal members of the public, our peers, graduate students, aspiring paleontologists, faculty, etc. This means our language is usually tailored to people with little background in paleontology, geology, and biology. We tend to include hyperlinks for extra information so if you don't need more information you can move past it but if you require clarification, you can still find it." One of the best features of the site is *Meet the*

Scientist, an interview series that features an astonishingly disparate array of scientists from many disciplines, include students, professionals at all levels, and amateurs, and represents a wide range of genders and ethnicities. It gives a resounding lie to the stereotypical images of what scientists are like, including, as Jen says, the "misconception that scientists are cold and calculated, which turns people away from wanting to interact or learn more. . . . This effort is working to increase visibility of scientists and showcase how diverse science and research can be."

Paleontology is media friendly. Dinosaurs and extinctions are not uncommon subjects of television documentaries, albeit of widely varying quality.[15] Kirk Johnson of the Smithsonian made an excellent three-part *Nova* series on the geological history of North America, *Making North America*. The shows based on Neil Shubin's *Your Inner Fish* are also a delight—Shubin's enthusiasm is infectious. And I give a resounding thumbs up to Emily Graslie's *Prehistoric Road Trip*.[16] In this fun three-part series, Graslie visits fossil localities (mostly dinosaur) and museums in the western United States and chats with the scientists working there. One of the remarkable things to me was the lack of the "usual suspects," those well-known male, white, and bearded paleontologists who have long been the public face of the field. This was a deliberate choice when she made the series; Graslie sought the more diverse voices she knew existed, including those of indigenous people. When I asked her about the focus on dinosaurs in the series, she responded that "dinosaurs are the charismatic megafauna! There is less emphasis on the rest of life; it is not as marketable. We also do a bad job teaching science that covers whole environments rather than just single organisms."

Graslie is not a paleontologist herself; she has a degree in fine arts. But she did an internship in natural history and came to the realization "that science is interconnected and requires imagination and creativity." As a student artist, Emily spent three years drawing simple white geometric objects under various settings. When she got to the museum and saw bones, she was very excited; she had a lot of questions about why they had the shapes they did. In art school, she also learned how people interact with art and how to grab their attention for more than thirty seconds. She parlayed this ability in her role as the Field Museum's chief curiosity correspondent, where she produced hundreds of episodes of the *Brain Scoop* series of YouTube videos. When I asked Emily what professional paleontologists needed to be doing, she mirrored comments made to me by Dena Smith: "We need to do a better job developing trust and empathy with diverse groups of people by being more respectful."

A critical area of outreach shown in *Prehistoric Road Trip* is public involvement in the process of fossil discovery. Many paleontologists rely on the help of volunteers in the field and the lab. For example, the Burpee Museum of Natural History in Rockford, Illinois, has for years brought members of the public out to their field sites in Utah and Montana to participate in the excitement of uncovering dinosaur remains.[17] An interesting example is the Homestead Site in Oklahoma. As described by Anne Weil, "a Board Member of the Sam Noble Museum had a long-cherished dream of providing a science education center for the youth of Oklahoma. His strongly held belief is that nothing will 'hook' someone on science faster than the thrill of discovering and excavating dinosaur bones. After much research and consultation with paleontologists, his foundation purchased land." At that time Anne was leading a field trip in the area of Native Explorers, an Oklahoma-based organization focused on recruiting and training Native Americans in the STEM disciplines. " I asked whether as part of the NE activities we could explore whether it was a viable site or not. We could. At the end of the day, we knew that it was. I then got permission from the landowner and the museum to run the site and persuaded the landowner to deed all fossils to the museum. The landowner's provision is, of course, that it be used as much for education as for research. As a result, the site is almost entirely excavated by Oklahoma students, from scouts to a couple of grad students." Preparation is mainly done by senior citizens in Tulsa!

A remarkable place the public can visit and see fossils "in their native habitat" is the Falls of the Ohio State Park in Clarksville, Indiana. Located along the banks of the Ohio River, the park exposes spectacular Devonian-age fossils beds, including some truly impressive ancient corals. Although the beds themselves are closed to collecting, the park staff has arranged for local quarries to donate piles of materials that can be combed through for fossils. There is also a summer Family Paleontology Camp and a newly renovated Interpretive Center. Alan Goldstein is the interpretive naturalist and park paleontologist and leads numerous on-site programs to the fossil beds. Alan has been involved in museum and park education since 1985, and his interest in fossils dates to early elementary school. His area of expertise is Devonian fossils and stratigraphy of the Ohio Valley and is centered around the Falls of the Ohio River.

I asked many of my colleagues what the public should know about paleontology and received a wide range of answers, and some general themes
emerged. First, they identify key concepts that are critical to know. The first
of these is deep time; as Pincelli Hull points out (and as the title of this book
seconds), "we are explorers of deep time." Emily Graslie placed deep time as a
concept central to the *Prehistoric Road Trip*, and Herb Meyer stresses that "it's
important that the work we do as paleontologists provides information and
new knowledge that helps the public to understand Earth's history."

The second general theme is evolution. Bruce MacFadden and Jacques
Gauthier agree that many members of the public lack a basic understanding of evolution. Bruce feels this is largely due to poor communication about
evolution. They also agree that museums are critical for explaining the science. Peter Wagner adds that people need to know how life on Earth got to be
where it is, with all the ups and downs over time, and how organisms adapt or
do not adapt to change. In particular, he believes it is critical for us to know
that replacing extinct forms is hard; it is easier to conserve what we have.

A third concept is that paleontology has valuable implications for environmental and climate science. As Arnie Miller put it, "What we work on has relevance to everyday stuff people care about." Susan Kidwell believes the public
should know that conservation paleontology exists; that our knowledge of
the younger parts of the fossil record can be brought to bear on conservation
issues. According to Rowan Lockwood, this is "applied paleontology"; the past
can tell us a lot about the impacts of climate change and extinction. Related to
this is the key role that fossils play in understanding ancient climates and in
predicting the responses to global warming and ocean acidification. According to Ellen Thomas, the use of fossils in "paleoceanography makes it possible
to reconstruct climate and environmental changes in the past, and thus [learn
about] the response of oceanic biota to the environmental changes."

My colleagues also identified a general need to address public misconceptions about who we are and what we study. There is frequent confusion between
paleontologists and archaeologists. Herb Meyer suggests that "if you're over
sixty and still don't know the difference between a paleontologist and archaeologist, it may be too late, but our staff will try to educate [you] anyway." Then
there is the assumption that paleontologists are only interested in dinosaurs.
Herb states that many of the people who visit Florissant National Monument
think " 'fossil beds' means dinosaurs, so broadening those perspectives helps

people to understand the many things that paleontologists can study." Ellen Currano agrees: "It's not just about dinosaurs. There are so many amazing questions that paleontologists are asking and using really diverse and innovative methods to answer them. I study plants because they are the base of all ecosystems on land—everything else depends on them. Plant fossils can be used to reconstruct past climate, to understand past plant life and vegetation structure, and fossil leaves show traces where the leaf was eaten by insects while still on the tree." There is also a lack of knowledge that we study what happened long before the dinosaurs, or even multicellular life, appeared. Sarah Tweedt wants the public to know how "much is still unknown about life and the world billions of years ago, but that we do know that it was weird compared to today." The fascinating question to her is, How did life arise on Earth? Sarah suggests that understanding this is the best way to understand the possibility of life on other planets; we should "send a paleontologist to Mars because they know what to look for."

A similar misunderstanding was pointed out by Lisa White, Mark Uhen, and Jess Theodor—we are only envisioned as being in the field collecting fossils. The public does not see us in the library, doing lab work, working with museum collections, writing grants, or training students. Susan Butts emphasizes how much of the job is repetitive lab work, gradually going through collections and picking out fossils. The public is also unaware of the diversity of our methods. As Stephen Dornbos puts it, paleontology "is a vibrant interdisciplinary science! We are not just stamp collectors. We use concepts and techniques from geology, biology, chemistry, and physics to reconstruct the history of life on Earth, and we get to travel the world while doing it." Lisa White concurs: "We do a bad job of explaining why we do what we do."

I believe a fuller understanding of the past is instrumental to our ability to solve the problems we face today. Paleontology provides critical insights into what happens to the biosphere when global environmental disasters happen. Paleontology provides vital information on the history of climate and environmental change that gives context to what is happening now. In addition, paleontology is a vital component in the search for fossil fuels and for determining the age of rocks. Paleontology is a great hook for the teaching of STEM. And paleontology provides direct evidence for the evolution of life on Earth. I can think of a no more integrative or important science.

19

THOSE WHO DO NOT KNOW THE PAST

The caller identified himself as the youth pastor of a church. After calling me by a nickname that no one except my pediatrician ever used (or has permission to), he asked if I could make a visit and talk to his young flock. About what? "The debate between godly creationism and god-less evolution." I ever so politely declined, saying something along the line of "I don't do debates." And I don't. As I explained to others later, a trained debater can effectively argue any side of an issue. I could have lost any debate, even if I were obviously correct.[1]

Paleontology has found itself somewhat central in two of the most contentious areas of the public perception of science and how it is taught. The first is the seemingly endless controversy over evolution. The other, of much more recent vintage but sharing many similarities in tactics and tone, is the argument over global climate change. Both debates have been covered at length elsewhere; I cannot do them justice here.[2] And I do not want to provide climate or evolution deniers any validation by discussing their ideas.[3] I will briefly discuss, however, some of the ongoing responses of the paleontological and broader scientific and faith communities to the challenges posed against science.

The most direct response is to present accurate science to the public and educators in a way that engages their interest. As I mentioned in chapter 5, the Paleontological Research Institution (PRI) and its beautiful Museum of the Earth have transitioned from a quiet research institution into a major player in STEM education and outreach. They have also moved from a focus solely on paleontology to adding a focus on climate change; evolution and climate change are now key topics in their exhibits, publications, and activities. Their main exhibit, *A Journey Through Time*, illustrates the history of life on Earth as documented by some choice fossils. In late 2020 they opened a new exhibit, *Changing Climate: Our Future, Our Choice*, with an accompanying online exhibit.[4] Where PRI really excels is in their educational materials. There is an online *Digital Atlas of Ancient Life*,[5] developed by Jonathan Hendricks, that includes a virtual collection of fossils and hundreds of 3-D models. It also contains a set of regional field guides. When the pandemic shut down in-person classes, many of us, me included, turned to the *Digital Atlas* to help us teach labs (thanks Jonathan!).

There is also a plethora of material on climate change. PRI operates the New York Climate Change Science Clearinghouse to help "support scientifically sound decision making in New York State." There are also numerous resources for teachers, including the very useful book, *The Teacher-Friendly Guide to Climate Change*, which can be downloaded for free.[6] This is part of a series of "Teacher-Friendly" guides that include books on evolution and on regional geology. The PRI education and outreach group, including Don Haas and Robert Ross, have done an amazing job.

Another resource with a great depth of material is the University of California Museum of Paleontology (UCMP). The UCMP site has long been the go-to for basic introductions to fossil groups and the history of the Earth. Under the leadership of Judy Scotchmoor, and more recently of Lisa White, they have developed incredibly useful webpages on both evolution and climate change that are notable not only for their content but also for their clarity and graphics. As billed, their website on *Understanding Evolution* is indeed a "one-stop source."[7] I know of no better place to learn the basics of evolutionary theory, as well as many of the details. I have used many of their figures in my own introductory classes. A very useful set of teaching materials, ranging from kindergarten to undergraduate, and an archive of resources on evolution are also available.

UCMP has recently released its *Understanding Global Change* website.[8] This site is outstanding for its interactive nature. For example, there is a highly detailed infographic showing the interactions of the causes of global change, their influences on the components of the Earth system, and impacts on the quality of human life. Clicking on any component or interaction on the diagram leads to an explanatory page. Another component of the website lets you construct a model for relevant processes and phenomena, such as habitat loss or greenhouse gas emissions. Again, this site incorporates a detailed set of teaching resources.

I should also mention a third website created by UCMP, *Understanding Science.* It is an excellent antidote for those who think the "scientific method" is the linear version we all learned in school (observation-hypothesis-experiment-theory). For those of us in the historical sciences, such as paleontology and geology, the classic physics or chemistry-based view of how science works is inappropriate—how can we do an experiment on an asteroid wiping out most life on Earth? The site demonstrates that the real process of science is far more complex, again using an interactive graph, which I use in my own teaching.[9] Understanding how science works is critical for dispelling the misconceptions that are rife in evolution and climate change denial.[10]

Since 1981, the National Center for Science Education (NCSE) has been central in the struggle by some against teaching evolution in the schools.[11] Eugenie Scott, its executive director from 1987 to 2014, became one of the nation's most familiar figures in the battle against creationism. NCSE led the legal effort in the 2005 case of *Kitzmiller v. Dover*, which resulted in the resounding decision that teaching the pseudoscientific idea of "intelligent design" was unconstitutional.[12] In 2012, the NCSE added fighting misinformation on climate change to their mission. The goal is to support "teachers gaining the skills, knowledge, and confidence to overcome their hesitation in teaching climate change and evolution." They also support teachers who find opposition to teaching these subjects in their communities and track and challenge legal and legislative efforts to undermine science education. I have been a proud member for many years.

The Paleontological Society (PS), along with virtually every other scientific society, adamantly supports evolution and its teaching. The PS has published a clear statement explaining their support: "Because evolution is fundamental to understanding both living and extinct organisms, it must be taught in public school science classes."[13] Similarly, the statement on evolution by the Society of Vertebrate Paleontology (SVP) is clear: "Evolution is fundamental to the teaching of good biology and geology, and the vertebrate fossil record is an excellent set of examples of the patterns and processes of evolution through time. We therefore urge the teaching of evolution as the only possible reflection of our science."[14]

One of the most persistent untruths is that people of faith cannot accept the reality of life's evolution. The PS statement is unequivocal about this: "This difference between science and religion does not mean that the two fields are incompatible. Many scientists who study evolution are religious, and many religious denominations have issued statements supporting evolution. Science and religion address different questions and employ different ways of knowing." This last statement is a direct reflection of the theme of one of Stephen Jay Gould's best-known books, *Rocks of Ages: Science and Religion in the Fullness of Life*, which he also expressed in an essay in *Natural History*: "each subject has a legitimate magisterium, or domain of teaching authority—and these magisteria do not overlap (the principle that I would like to designate as NOMA, or 'nonoverlapping magisteria')."[15]

The PS has also promoted the compatibility of science and religion through its Distinguished Lecturer program. Longtime Distinguished Lecturer Patricia Kelley (box 19.1), former president of the society, is a person of faith and the wife of a Presbyterian pastor. She has taught Sunday school and Bible study. One of her Distinguished Lecturer programs is on "Evolution and Creation: Conflicting or Compatible?" where she seeks to bridge the divide between religious traditions and science. Writing in *Geotimes* in 2000, Tricia discusses how she interacts with students with creationist beliefs. She emphasizes that most issues come from student's misunderstandings of science and how it differs from religion, and that they need not conflict; science does not threaten faith. She also stresses the importance of being respectful of student's religious beliefs, while still expecting "them to know the evidence that has been used to support evolution, as well as the current explanations for how evolution occurs."[16]

BOX 19.1

Patricia Kelley, recently retired from University of North Carolina Wilmington, had an early interest in art. "My parents indulged my early interest in dinosaurs, giving me a beautifully illustrated dinosaur book for Christmas when I was seven. I loved to paint, so immediately began to copy the illustrations in the book. Coincidentally, about this time a new dinosaur hall opened at the Cleveland Museum of Natural History and my obnoxiously proud grandmother made me drag my paintings down to the Museum on opening day, where we showed the pictures to Museum director Bill Scheele. To my utter amazement, he commissioned more paintings and put them on display in the Museum for seven months!" However, *"like so many young women, I forgot science by the time I reached high school."* Tricia went to the College of *Wooster, where her randomly assigned advisor was paleontologist Richard Osgood, who suggested she take geology as an elective. Her interest in fossils was reborn. She went on to study with Steve Gould at Harvard for her PhD (unlike me, she did not turn down the opportunity).*

One key point that Kelley makes is that most mainline religious groups support evolution and oppose the teaching of creationism, in any of its many guises. An entity that works hard to support this reality is the Clergy Letter Project.[17] Founded in 2004 by Michael Zimmerman, the heart of the project is the *Clergy Letter—from American Christian Clergy—An Open Letter Concerning Religion and Science*, which states in part: "We the undersigned, Christian clergy from many different traditions, believe that the timeless truths of the Bible and the discoveries of modern science may comfortably coexist. We believe that the theory of evolution is a foundational scientific truth, one that has stood up to rigorous scrutiny and upon which much of human knowledge and achievement rests. . . . We ask that science remain science and that religion remain religion, two very different, but complementary, forms of truth." As of the end of October 2020, it had received 15,647 signatures. There are separate letters for Jewish, Buddhist, and Humanist clergy and a more recent one on the climate crisis. Beginning in 2006, the project also organizes Evolution Weekend, where congregations discuss issues at the intersection of science and religion.[18] In 2020, 232 congregations participated. There is also a list of scientists who are willing to work with clergy and to help with questions they may have. I am on that list.

Another organization that attempts to bridge the science-religion divide, from a predominantly Jewish perspective, is Sinai and Synapses.[19] This organization sees that "in order to enhance ourselves and our world, we need both religion and science as sources of wisdom, as the spark for new questions, and as inspiration and motivation." They proactively try to engage students and clergy with each other, to give a voice to those who have not usually been heard in the public battle between religious fundamentalists and adamant atheists.[20] I have participated in their Scientists in Synagogues initiative, where scientists are encouraged to interact with the congregations for discussions on Judaism and science. Along with two physicists and our rabbis, we discussed the nature of time in science and faith. I took the opportunity to discuss what we knew of the origins and evolution of the Earth and life, and the eventual fate of our planet.

Global climate change is real and threatens human society; evolution is one of the best supported concepts in science, there is no doubt life has evolved. These are not matters of belief; to quote a fellow Bronx boy (and fellow Bronx Science grad) Neil deGrasse Tyson, they are "true whether or not you believe in it." But we err in assuming the hostility of people of faith in these scientific concepts. They are more often than not key allies.

20

DRAWING ON THE PAST

As a child, I read science books voraciously, especially ones about astronomy. The one about paleontology that sticks in my memory is the *Giant Golden Book of Dinosaurs and Other Prehistoric Reptiles*, published in 1960, written by Jane Watson and illustrated by Rudolph F. Zallinger. On the cover, *Brachiosaurus* sits in a swamp with its mouth full of vegetation. In the background, an *Allosaurus* walks erect, its tails dragging on the ground, and a pterosaur flies by in the sky. Inside, another *Allosaurus* is chomping on the neck of a hapless *Brontosaurus*, and a *Tyrannosaurus* is facing off with *Triceratops*. None of the animals has feathers or any other body coverings. For those who grew up in 1950s, 1960s, and 1970s, the portrayal of dinosaurs by Zallinger colored our picture of what dinosaurs looked like. His magnificent mural, *The Age of Reptiles*, is a highlight of the Great Hall of the Yale Peabody Museum. A version appeared in *Life Magazine*. Perhaps only his predecessor, Charles R. Knight, whose murals hang in many American museums, including the American Museum of Natural History and the Field Museum, had as much impact on the public perception of prehistoric life.[1]

Today, the Zallinger and Knight murals are more notable as examples of superb art than as accurate portrayals of the ancient world. Even the venerable *Golden Book* has been supplanted, with a 2013 edition written by Robert Bakker and illustrated by noted artist Luis V. Rey. On the flyleaf, Bakker notes that the *Age of Reptiles* mural is what got him into science. The composition of the cover image had not changed much, but the details reflect current thinking about the appearance and biology of the animals. *Brachiosaurus* stands on dry land; *Allosaurus* leans forward, its body nearly horizontal. And inside the book, feathers cover nearly all of the dinosaurs. These changed perspectives on dinosaurs, as embodied by the artwork, is now the standard for children's book on dinosaurs. This generation will grow up with a very different mindset! This transition highlights the key role images of fossils have in influencing our mental images of the past.

It is very hard to separate fossils from the art used to portray them. Part of the training of every paleontologist is to learn to draw the objects they are looking at. When I was an undergraduate working at the American Museum of Natural History, the late Norman Newell told me that the very first step in understanding a fossil was to draw it. Drawing forces you to look carefully at what is there. My PhD advisor, Jack Sepkoski, said something similar about statistical analyses; his "Zero Rule" was "draw a picture first." I must admit that I am better at drawing graphs than fossils; at least I could use a computer to do the former. One useful crutch to someone with my limited artistic skills is a wonderful device known as a "camera lucida." This equipment attaches to a microscope and lets you see the specimen and the drawing at the same time, so you can trace what you see. Many paleontologists, such as Derek Briggs, use camera lucida drawings as a valuable supplement to photographs. In my case, it is much better than a freehand drawing.

A surprising number of paleontologists started out as artists or have adapted it as a hobby. Paul Sereno, a University of Chicago dinosaur paleontologist, majored in art and biology. Lisa White started out as a photography major, and Dena Smith began as an art major with a specialization in art illustration. The Smithsonian director Kirk Johnson has a bachelor's degree in geology and fine art, and Lisa Park Boush paints.

A paleontologist's ability to draw can be an essential contribution to conveying scientific ideas. John Ostrom's classic paper describing *Deinonychus* contained Bakker's oft-reproduced drawing of it running. This figure, along with the illustrations in Bakker's own book, *Dinosaur Heresies*, strongly influenced the mental images that accompanied the "dinosaur renaissance" begun by Ostrom. To the paleoartist Luis V. Rey, this was a transformation in "how to look at dinosaurs not as monuments in museums but as real animals." Any reader of a paper or book by Adolf (Dolf) Seilacher quickly realizes that his line drawings were there to convey the core of his argument and to convince you of its correctness. His drawing style is recognizable at first glance and was integral to his approach to science. In his acceptance speech for the Paleontological Society Medal in 1994, Dolf listed ten ideas as his credo; the third was "the value of self-drawn illustrations for the observational process."[2]

Fossils themselves have aesthetic value. Like shells and butterflies, they are collected and displayed as much for their intrinsic beauty as for their scientific interest. Fossils ranging from trilobites to dinosaurs are purchased much as one would buy a sculpture or a painting to decorate a home. I have seen large ammonites and slabs of petrified wood made into coffee tables. One company lets you tile your kitchen or bathroom in tiles made from Green River shale, each with a fossil fish in the middle of it. At the extreme end, some spectacular fossils can go for huge amounts at auction. In 2014, a palm frond from the Green River Formation was auctioned by Sotheby's for about $108,000. At the same auction, a complete mammoth skull with tusks from Siberia netted nearly $180,000. Needless to say, the purchase and sale of fossils as objets d'art is not without considerable controversy and is part of the complicated issue of the sale of fossils and commercial collecting (see chapter 16).

One fascinating take on fossils as art was the exhibit *Fossil Art*, put together in Germany by Dolf Seilacher. In this case, the "artists" were not modern humans but ancient animals as they moved across and through the seafloor, "drawing" their paths in the sediment. Many of these trace fossils are quite complex and ornate, with branches and swirls that remind one of abstract art. Working with a preparator, Seilacher made large latex molds of the bedding surfaces in the field. This allowed the material to be replicated and transported, as well as duplicated without damaging the original material. In his foreword to the exhibit catalog, Stephen Jay Gould wrote: "In the best traditions of melding art and science, these objects are both lovely to the eye, and inspirational to the brain."[3]

Also lovely in their own way are the drawings made to illustrate fossils, especially before the onset of photography and the ability to easily reproduce photographs in print.[4] Many of the old volumes on fossils, such as the thirteen folios of the Paleontology of New York or monographs of other state and national surveys, are illustrated with carefully prepared drawings of the fossils. One can often find these plates removed from the volumes for separate sale (I admit that I have a few on my own wall).

We can roughly identify three classes of images of fossils and ancient life, differing in their degree of interpretation of the original fossils. Closest to the original specimens are the illustrations prepared for technical publications. The goal of these is to supplement and enhance the written descriptions and to clarify the features that may be unclear in photographs. Although somewhat interpretative, these drawings must be as accurate as possible because they are often used by other paleontologists as references and comparisons.[5] As noted earlier, the advent of digital imaging and related methods have profoundly changed how we visualize and describe fossils. Although line drawing is still a critical component for many published descriptions, the preparation of even these has been greatly assisted by computer technology. I cannot imagine making figures for a paper without using technological aids.

Many fossil remains are fragmented, scattered, incomplete, distorted, flattened, etc. Based on their knowledge of anatomy and morphology, paleontologists attempt to envision what a complete organism would have looked like, or at least a complete skeleton in the case of vertebrates. The result is a reconstruction of the original appearance of the remains prior to the vagaries of fossilization.

Producing a reconstruction is where the artistic abilities (if they have any) of paleontologists are brought into play. For those of us who lack artistic ability, this is where we first start to interact with artists. Beginning in 1975, Desmond (Des) Collins of the Royal Ontario Museum led eighteen expeditions to the iconic Cambrian-age Burgess Shale of British Columbia. During the 1983 expedition, they found a remarkable new arthropod in the beds at Mt. Stephen. In a paper by Collins and Derek Briggs. they named it *Sanctacaris* ("saint crab;" informally Santa Claws) and recognized it as an early member of the group that contains modern scorpions, spiders, and horseshoe crabs.[6] To illustrate the paper, Des approached Marianne Collins, a museum artist who specialized in wildlife art; this was her first fossil reconstruction for a research paper. Marianne describes her goal as "to represent the animal as true to life as the science can show." Since her original reconstruction of *Sanctacaris*, she has illustrated more than eighty species of Burgess Shale fossils, to the extent that "I feel that they are family. Many of these fossils are new discoveries or need redescription due to new findings." A lot of us are familiar with Collins's art through her illustrations in Stephen Jay Gould's bestseller, *Wonderful Life*. This book and its figures were paleoartist Barbara Page's introduction to paleontology: "The illustrations of reconstructed organisms were clear and concise while the actual specimens often required a trained eye to discern."

Because reconstructions are interpretations, they can change dramatically as scientific knowledge and opinion changes. A familiar example is the change in how the skeletons of dinosaurs are illustrated and mounted in museums since the dinosaur revolution. No more tails dragging on the ground! In a 1996 paper, Des Collins showed how reconstructions of the iconic Burgess Shale predator *Anomalocaris* have changed over time, based largely on new specimens.[7] In my own work, an oft-republished reconstruction of the Mazon Creek animal *Essexella* showed it to be a jellyfish; we now believe it was reconstructed upside-down and was in fact a sea anemone. Same specimens but a different way of understanding them.

Reconstructions, especially when done by or in cooperation with an expert paleontologist, are interpretive but still attempt to hew closely to the preserved material. Reconstructions also show the plant or animal isolated from their biological and physical environment. It is when an attempt is made to visualize the organism in the context of its wider world that reconstructions truly become paleontological art (paleoart). In 2014, paleontologists and the paleoartists Mark Witton, Darren Naish, and John Conway defined paleoart as "a branch of natural history art dedicated to the reconstruction of extinct life."[8] To Marianne Collins, " 'illustration' is where the artist tries to represent what is and 'art' is where the artist is free to interpret reality."

Paleoart results from the creativity of the artist channeling the knowledge of the scientist. The very best paleoart is not only beautiful but is also scientifically accurate. In the later 1970s, William Stout prepared forty-four drawings for a revision of writer and dinosaur enthusiast Don Glut's popular book *Dinosaur Dictionary*: "As I was drawing them, I thought, 'This may be the only picture the public ever sees of this animal, so it had better be accurate.' " The vertebrate paleontologist Darren Naish praises paleoartists "who pay close attention to what we really know of soft tissue anatomy, who try to be rigorous in choosing what to depict and how, and who make special effort to depict ancient animals and environments as biologically plausible."

What is represented in paleoart is not necessarily "truth." As aptly put by artist Wendy Luck, "paleoart will by necessity include assumptions and misconceptions held by the artist. It is art from our imagination after all, even when based in the best science we have." Evan Sizemore adds: "Of course, by its very definition, all paleoart is inherently wrong—a lie, even. No one will ever perfectly reconstruct an extinct animal that lived before the dawn of man. But that also means that I can have a direct hand in recreating these beasts—I decide how fat or skinny they will be, what colors they don, stripes or spots, etc."

I was struck by the variety of artistic expressions based on the same fossils at a 2017 exhibit at the Yale Peabody Museum about the iconic ancestral bird *Archaeopteryx*.[9] The fossils of this animal are incredibly rare, and the exhibit brought together casts of almost all of them. It also presented artistic conceptions of *Archaeopteryx* by six noted paleoartists, including Luis V. Rey, Gary Staab, and William Stout. A mural produced by Rey showed not only *Archaeopteryx* but many of its relatives in their feathered glory. My niece's husband stared at the mural and told me "that is not at all how I pictured dinosaurs." One more person educated.

Much of this variation in how paleoartists represent fossils is grounded in their personal style, including the media they use. One extreme is photorealism, where an attempt is made to create an image that resembles a photo as closely as possible, to the extent that it may be difficult to tell that the artwork is not a photo or a movie of living things. The moving images of *Tyrannosaurus* at the Field Museum's recently reopened SUE exhibit is one such example, as are the dinosaurs of Jurassic Park (that look real but are inaccurate). I am especially fond of the Burgess Shale films made by the apparently defunct Phlesch Bubble Productions, which give a real sense of being there. The reconstructions of fossil hominids by John Gurche look like you could walk up and try to talk to them.[10] Gary Raham says that he enjoys photography and "mostly works in a photo-realism style," but he also does "a lot of graphic art, which is more stylized." Similarly, Luis V. Rey indicates that "these days my reconstructions are a bit more graphic, almost photographic . . . but not merely photographic. The technique should never overcome style." Jimi Catanzaro describes his work as " 'stylized realism', meaning that everything is anatomically correct, scientifically accurate, and there is definitely a sense of realism to the drawings, however the plan is usually to put emphasis on specific parts of the composition, leaving some elements simplified or stylized rather than achieving a photo-realistic quality to the drawing."

Of course, not all paleoart is photographic or attempts to be. One of my favorite paleoartists is Alaska-based Ray Troll.[11] His artwork combines anatomically correct living and fossil animals with elements of whimsy and surrealism; his work is not photorealistic but is designed for both artistic and informational impact. Many of us own a delightful T-shirt design he did for the Paleontological Society illustrating the geologic time scale.

By his own admission, the work of William Stout is definitely not photographic, nor is he a fan of that style of art, except for the work of John Gurche and a few others: "I find it both boring and a lie. That tight photographic style tends to assume we have more knowledge than we actually have. I also find that kind of work to be very tedious. I think art should be playful and fun." Stout's favored media are oil on canvas, pen and ink, and pen and ink with watercolor, which he points out "won't give you a photographic style."

The media that paleoartists work with are as varied as their styles. Many work with multiple media. Barbara Page uses "almost anything that suits the conceptions I cook up in my head: watercolor, ink on paper, acrylic and oil paint, plus collage." Gary Raham, like Stout, works in pen and ink, but he mainly uses pencil and acrylic. Pencil and paper are also favored by Evan Sizemore, whereas Jimi Catanzaro enjoys working with ink. Marianne Collins also worked in pen and ink, in addition to Conté sticks, acrylic sticks, gouache, and watercolor. She also makes three-dimensional urethane models, which she then paints. Gary Staab is a specialist in making models, not only for paleontology but for a wide range of natural history subjects. He has created sculptures and models for museums all over the world. Staab likes to joke that "I specialize in being a generalist."

Similar to other aspects of science, the advent of fast and convenient computers has revolutionized paleoart. "Since the advent of computer paint programs, most of my illustration work is digital," says Marianne Collins. Many, such as Gary Raham and Luis V. Rey, use programs such as Illustrator and Photoshop. Jimi Catanzaro employs Photoshop when a project requires color.

It is not surprising that dinosaurs are by far the favorite subject for paleoartists (and blockbuster movies). Nearly all the posts by the 12,500(!) members of the paleoart Facebook groups are of dinosaurs. Like Wendy Luck, they prefer dinosaurs because they never outgrew their "childhood love of them."

As Luis V. Rey puts it, "dinosaurs exert a fascination that is difficult to overcome." Evan Sizemore admits a "bias toward tetrapods, and reptiles in particular, especially dinosaurs"; and William Stout states, "obviously, I've painted lots of dinosaurs." Ray Troll has been drawing dinosaurs all his life, from his earliest memories of drawing with a crayon at age three to a just completed drawing of a *Hadrosaurus* on the wall of the Academy of Natural Sciences in Philadelphia. What beguiled him was the idea that dinosaurs were real.

Evidently, experienced paleoartists go way beyond dinosaurs. For example, Rachel Rasfeld grew up in the Cincinnati area, with its widespread and accessible Ordovician fossils, including trilobites. One of them was the star of her childhood fossil collection. She retains her love of trilobites to this day: "trilobites might not be featured in a Jurassic Park–style blockbuster, but to me they are just as amazing as the dinosaurs. I want other people to feel that too, even just for a moment." Gary Raham has concentrated "on fossils that I can readily collect," which includes Paleozoic marine fossils, such as trilobites, from Utah and Colorado. William Stout has painted "nearly every type of prehistoric creature."

What often determines what a paleoartist will create results from a close interaction with professional paleontologists. I first interacted with Ray Troll when he asked me about eurypterids for a book he was illustrating. At an exhibition at the Burke Museum in Seattle, Troll had a fateful meeting with Kirk Johnson, a paleontologist then at the Denver Museum of Nature and Science and now director of the Smithsonian National Museum of Natural History. The two of them ended up taking a road trip together through the Rockies that they chronicled in the 2007 book *Cruisin' the Fossil Freeway*, written by Johnson and illustrated by Troll. A sequel, *Cruisin' The Fossil Coastline*, came out in 2018.

Troll has worked with many other paleontologists, especially those who study fish, and in particular sharks, his other great artistic passion. As he puts it, "Kirk Johnson says he is a scientist who collect artists. I am an artist who collects scientists." One of his recent collaborations was with Leif Tapanila, director of the Idaho Museum of Natural History. Tapanila used new methods, in particular CT-scanning, to reexamine museum material of the unique Permian "whorl-toothed shark" *Helicoprion*. Troll provided important input to the entire project, which indicated that the animal was related to ratfish and not sharks, and he drew the reconstruction that illustrated the paper.[12] Troll's illustration also became a major component of a traveling exhibit about *Helicoprion*, which also includes a life-sized model by Gary Staab.

Although they generally lack formal training themselves, all the paleoartists I interviewed work closely with paleontologists, and those early in their careers wish they can. Luis V. Rey calls professionals "sources that feed anything that I do in the field of paleontology." This is clearly a synergistic relationship; William Stout is "not happy with a reconstruction until the paleontologist I am consulting with is happy, too, and has no reservations about anything within my picture. When doing a reconstruction, I typically try to work with an expert on that animal or the paleontologist who discovered it."

One fascinating project arose from Barbara Page's interactions with Warren Allmon, director of the Paleontological Research Institution. Allmon was in the planning stages for what would become the Museum of the Earth and wanted an exhibition that provided a visual journey of the evolution of life. Artist Barbara Page was doing research in the library at PRI for a one of a kind book illustrating the passage of time when Allmon saw her paintings of fossils in their matrix. He proposed that she do a mural for the Museum of the Earth, then in the planning stages. "Ultimately . . . the project grew into a permanent installation of 543 contiguous painted panels, each of which represents a million years in the history of visible life." You can best see Page's remarkable artwork at the Museum of the Earth (more than five hundred feet of bas relief panels!) or alternatively in the accompanying book, *Rock of Ages, Sands of Time*.[13]

Of course, Page did not do her work for free or to "get exposure." Like any artist she deserved and was paid a commission. One would hope that all paleoartists could make a living from their art. Some do; Luis V. Rey has always tried "to live from publishing my artwork," including recently from his own books, but adds: "If you think you are going to get rich via "paleo art" you are in the wrong business!" William Stout has a similar view: "I make a very good living from my art—but I'd starve if I had to solely depend upon paleoart for a living." His work has appeared over the last fifty years "in film, on comics, book illustrations, movie posters, theme park designs, T-shirt designs—nearly every aspect there is in the field of art." Ray Troll has a physical store in Ketchikan, Alaska, and has a strong online presence. Others work outside of paleoart to support themselves; Gary Raham is employed as a graphic artist, and Wendy Luck is a program manager. Jimi Catanzaro's full-time job is as a technical illustrator.

Like everyone in the creative field, the livelihood of paleoartists is threatened by the internet and the culture of copying, which has been made especially easy by the advent of digital art. This can either involve using one's works without permission or creating derivative works that closely mimic the original artist's conception. The latter, as pointed out by Witton and colleagues in their insightful essay, leads to the variations of the same imagery, often inaccurate, being seen again and again. These memes within paleoart demonstrate "the problems of decreasing accuracy introduced into palaeoart through copying work, as well as a deficit of originality."[14] They suggested that those who commission it go by the slogan "Support Original Palaeoart."

Nearly all the paleoartists I have spoken to have had their images reproduced without their permission to the extent that it sounds like a given that it will happen. Marianne Collins "found my own original pen and ink drawings colorized by another artist, another copied down to the pose and used in a museum exhibit or popular science book and even sold on museum gift store merchandise. And the public doesn't know the difference, so mistakes and misrepresentations are perpetuated." "It's especially rough when your goal is to make a living by selling your art, and another person [is] benefiting financially by stealing your work. But all you can really do in that situation is pick up and move on, try to be as diligent in protecting your work as possible without letting yourself get discouraged, or allowing yourself to forget why you are doing this in the first place," says Jimi Catanzaro. A more proactive stance is shared by William Stout, who has "to pursue every misuse and violation. If I don't, I risk losing my copyrights."

When professional paleontologists and paleoartists work together, the result is a much-improved understanding of life of the past. Paleoartists are also our partners in presenting our science to the public. As Evan Sizemore says, "when artists both do their homework and take (realistic) creative license, it can be the mental push that both paleontologists and the public need to start seeing these animals in a more dynamic fashion, to break out of the box or to start asking questions they hadn't considered before." From Darren Naish: "Our knowledge of ancient life is driven by discoveries in the field, by the excellent science done in various laboratories and research centers etc. BUT the way we imagine these creatures and their worlds, what we think about specific aspects of their posture, anatomy, and behaviour is led by a cadre of artist-researchers who are not just presenting the 'face' of palaeontology, they are also combining numerous strands of evidence to show and remind everyone else interested what is deemed plausible and valid right now."

I will give the last word to Rachel Rasfeld: "Art and science have much in common and much to offer one another. Paleontology is an excellent gateway to the natural world in general. The allure and mystery of long-extinct living things captures the imagination, and such curiosity inevitably leads to a better understanding and appreciation of both life lost and the life on our planet now."

21

THE WORLD OF PALEONTOLOGY

n the fall of 1984, with the ink still wet on my doctoral diploma, I traveled to Edinburgh, Scotland, to attend a conference on *Fossil Arthropods as Living Animals* at the Royal Society of Edinburgh. There, for the first time, I met the many senior paleontologists from around the world who worked on extinct arthropod groups. Understandably nervous, I presented my research suggesting that eurypterids, like modern blue crabs, used their appendages to generate lift, "flying" under water. To make my interpretation, I had used the superb detailed reconstructions of eurypterid appendages by Paul Selden, who had alternatively concluded that they used them to generate drag—"rowing" instead of flying (Paul was there). At the end of the day's events, just as I began to relax, the conference chairman announced that the next day would be devoted to the discussion of issues, such as the "disagreement between Dr. Selden and Dr. Plotnick." So much for the planned relaxing evening with my wife.[1]

This was my first introduction to the international conference and to the truly global scope of paleontological research. With the notable exception of the amazing German born Curt Teichert, all my professors were American.[2] Although I had met foreign scholars when they visited my graduate schools (Adolf Seilacher is one memorable example, as he was to all paleontologists of my generation), nearly all of my contacts had been with English-speaking, predominantly North American (United States and Canada) scholars. Even to this day, I am most comfortable and familiar with American paleontology and paleontologists. You may have noticed that when reading this book. This is unfortunate because there is no American exceptionalism in paleontology; research in other countries equals and often exceeds the quality and certainly the quantity of what is done here. And many of the challenges—insufficient funding, a lack of positions, a loss of expertise in specimen-based paleontology, and the illegal removal of fossils—are worldwide issues. The last of these issues has grown as nations act to treat fossils as parts of their own cultural heritage and view their removal as a vestige of colonialism. In this chapter, I touch on the status of my discipline in countries and regions across the globe.[3] It is by no means comprehensive or representative of the work done in almost every nation.

As mentioned earlier, some of the world's key fossil sites, such as the Burgess Shale and Mistaken Point, are in Canada. The Royal Ontario Museum in Toronto houses one of the major collections of fossils from the Burgess Shale and is in charge of ongoing excavations there. Western Canada also has some of the most important dinosaur sites, with world-class collections at the Royal Tyrrell Museum of Palaeontology. Many Canadian universities have strong paleontology programs, and there are close ties between paleontologists in United States and Canada. (My most recent PhD student, Shannon Hsieh, is from Toronto, and I have coauthored papers with Canadian researchers.)

Although people across the world have been familiar with fossils for millennia, the intellectual roots of the modern field are in Europe,[4] and Europe remains central to the growth of paleontological knowledge to this day. World-renowned research is conducted in Italy, Spain, Portugal, Ireland, Russia, Poland, Sweden, Norway, and the Netherlands as well, but I will regrettably restrict my discussion to Germany, France, and the United Kingdom.

As would be expected from their almost common language, there are close ties between paleontologists in the United States and Canada and those from the United Kingdom and Ireland. Many paleontologists at North American universities and museums are British or received their training in universities in the UK. At the same time, American-born or -trained scientists have received positions in the UK. This has resulted in what is effectively a single research community. For example, I have written numerous papers with Andrew Scott of Royal Holloway, University of London.

Viewed from this side of the pond, British paleontology is dynamic and cutting edge. Many universities in the UK have active paleontology programs and produce high-quality students. And there are a lot of museums with significant fossil collections; not just the venerable Natural History Museum in London (which is a must visit). A 2016 article in *The Guardian* listed nearly three dozen such museums throughout the UK and suggested that their list was incomplete.[5] (A visit to Lyme Regis to see the museum there and collect where Mary Anning did is on my bucket list.)

The lead British paleontological organization is the Palaeontological Association (PalAss), founded in 1957. It is an international organization, whose one thousand members come from about fifty countries. Like the Paleontological Society, it publishes two journals, with *Palaeontology* focusing more on the paleobiological aspects of the field and *Papers in Palaeontology* concentrating on descriptive paleontology; it also produces a very impressive newsletter. The PalAss also holds annual paleontology-oriented meetings, including Progressive Palaeontology, where graduate students[6] are given an opportunity to present their research. I recently attended (virtually) a PalAss annual meeting. One key area of discussion was how to deal with the legacy of colonialism in museums.[7]

Some of the same social and legal issues that concern American paleontologists also occur in the UK. British collecting rules are as complicated as those in the United States.[8] In a 2018 survey, Richard Butler and Susannah Maidment examined the long-term career prospects of doctoral students.[9] They focused on those who presented as graduate students at the Progressive Palaeontology meetings from 2000 to 2008. They found that 44 percent were still in academia, although not necessarily in paleontology. What is disturbing, however, is the gender disparity: 61 percent of the male but only 20 percent of the female students remain in academia. Similarly, a 2020 article found that less than 20 percent of the authors in the journal *Palaeontology* were female and that that number had not changed in two decades.[10] Finally, until recently British scientists had benefited from easy travel to, access to facilities, and available financial support from the European Union. Brexit has put the future of these benefits in doubt.

Paleontology in the UK follows in the footsteps of Hutton, Lyell, and Darwin. In France, it follows in the footsteps of Cuvier and Lamarck. The history and status of French paleontology was thoroughly reviewed in 2012 by Thomas Servais and his coauthors.[11] The most important vertebrate collections are in Muséum National d'Histoire Naturelle in Paris, which is also home to the largest group of research paleontologists in France. The major invertebrate collection is at the University of Lyon 1. Servais and colleagues state that "it is difficult to explain the organization of paleontological research," it being a mix of affiliations with universities and with the Centre National de la Recherche Scientifique, a national unit that supports scientific research. They estimated that there were about 150 to 250 paleontologists in France. Many of these belong to a younger generation of scientists that became established in the early 2000s. This generational change was mirrored in a shift in focus, as it was in the United States and Britain, away from descriptive studies and geology and toward broader studies in areas such as climate change and paleobiology. And, as is the case globally, these scientists now tend to publish in English, even in French published journals such as *Comptes Rendus Palevol.* The professional society, the Association Paléontologique Française, is relatively young, founded in 1979, and has about 250 members. Servais et al. identify a key issue for French paleontology as the lack of "strong academic visibility. . . . This new generation of scientists must find a unity and create lobbying organisms to ensure financial support for paleontological studies in the new very competitive funding environment."

German paleontology has also long been a powerhouse. Dolf Seilacher was the face of German paleontology to many of us in the United States, but he is part of a long intellectual tradition going back to the eighteenth century and continuing today. German paleontologists have many obvious strengths: they conduct highly innovative and high-impact research; Germany has many key fossils sites; and some of the very best museums in the world are in Germany, such as the Museum für Naturkunde in Berlin (also a must visit) and the Senckenberg Museum in Frankfurt. German paleontology is also relatively well-funded. Unlike trends in the United States, there is still a strong foundation of specimen-based research. There are several high-quality paleontology journals, including the descriptive paleontology-oriented *Paläontologische Zeitschrift* (now known as *PalZ;* and yes, its papers are now in English). *PalZ* is published by the large professional paleontological society, the Paläontologische Gesellschaft.

Despite these obvious strengths, a 2010 essay by Wolfgang Kiessling (Universität Erlangen-Nürnberg) and coauthors raised some troubling developments.[12] Many of the concerns they express should be familiar to American paleontologists, such as the closing of departments and the erosion of positions, with paleontologists being replaced by nonpaleontologists (by 2019 Germany had fewer than fifty professors of paleontology). They also recognized the image problem that paleontologists are "people who dig for bones in remote areas but otherwise contribute little to the advancement of science." The essay includes positive suggestions for the future, such as, as their French colleagues suggested, lobbying to find new allies that recognize that "only we have the ability to study ecosystems undamaged by human activities and their response to natural physicochemical perturbations." One point they made was that German paleontologists have been slow to move away from strictly geological research and training and toward a better integration with biology. This is now being addressed. Kiessling has told me that "the situation for paleo has improved because our discipline has changed to become more relevant and is on a good track." Especially exciting is a large grant to Kiessling and Manuel Steinbauer from the Volkswagen Foundation to support the Paleo-Synthesis[13] project, which began in late 2019. Supported by an international board of directors, the project will hold international workshops "to work on research questions that arise from a joint discussion of the paleontological community," run training classes on quantitative methods and data analysis, and have a strong outreach component, including citizen science. Very exciting developments!

Thomas Servais has been president of the *Association Paléontologique Française* and vice president of the Palaeontogical Association and of the International Paleontological Association, which is a global umbrella federation.[14] He is not as sanguine as Kiessling on the future of paleontology in Europe. Servais is deeply concerned about the obsession of universities with the H-index and other measures of the impact of papers and journals (see chapter 13) and its impact on research. This is reflected in the topics researchers choose to work on ("everybody is working either on the Cambrian explosion or on dinosaurs"), the time periods and groups studied ("why work on Carboniferous brachiopods or Cretaceous rudists,[15] when you get much more attention, because the community is much larger, with Quaternary foraminifers?"), and the fields that are hired ("in France and Germany, in the two last decades, palaeontologists have been massively replaced by geoscientists or bioscientists that are working in disciplines with higher publication impact; in France, many jobs went to isotope geochemistry"). Servais also bemoans the lack of knowledge of the fossils themselves (classical paleontology) by either isotope geochemists who "study brachiopods or conodonts they are unable to identify at the species level" or by paleontology PhD students who "work now on databases but are unable to identify the fossils that are in the data sets!" As a palynologist (fossil plant spore specialist), he is also concerned with the decline in the community working on anything older than the Quaternary. Servais sees the same issue in paleontology as a whole.

Over the last two decades, China has rapidly risen to become a world leader in many areas of paleontological research.[16] New discoveries from Chinese localities appear seemingly weekly in the pages of *Science* and *Nature*. These include the amazing Cambrian site of Chengjiang, which equals or surpasses the Burgess Shale and the Cretaceous-age Jehol Biota, which has yielded remarkably preserved feathered dinosaurs and a host of other well-preserved fossils. These sites, and others, have literally transformed our perspective on the evolution of life on Earth, including the origins of our own group, the mammals.[17] Chinese paleontologists have become fixtures at international meetings. In a preconvention field trip for the 2019 North American Paleontological Convention in which I participated, half of the participants came from Northwest University (China), a public university in Shaanxi Province. This also reflects that paleontological research is well funded in China. Chinese scientists have become frequent contributors to English-language journals. This is very clear, for example, in recent issues of the major international journal *Palaeogeography, Palaeoclimatology, Palaeoecology*. One recent change I have noted is that the number of Western coauthors has dropped, showing greater confidence by the Chinese in taking leadership roles.

Although specimen-based research still dominates, with rising young stars, Shuhai Xiao (Virginia Tech) indicates that work in quantitative paleobiology "has improved significantly compared to 10 years ago. A large project (Deep-time Digital Earth) was initiated a few years ago, and the GeoBiodiversity Database or GBDB[18] is a major component of the DDE project. This year the GBDB published its first significant results in *Science*. GBDB is complementary to the Paleobiology Database (PBDB), and it is a positive step forward toward addressing the weakness in quantitative paleobiology in

China."[19] Zhe-Xi Luo (University of Chicago) agrees and points out that the "information content of GBDB is more fine-grained and precise than that of PBDB," although the PBDB is open-access and more user friendly. Luo indicates that China has had some success in analyzing paleontological data, but China still lags behind the United States and Europe. He also points out that strides have been made in ancient DNA (aDNA) studies. Xiao and Luo also remarked on markedly improved training in paleontology at both the graduate and undergraduate level, with reputations for scientific excellence spreading beyond the big institutes and universities. As an American paleontologist, all I can be is jealous.

In a 2010 review of the status of paleontology in China, Xiao, Qun Yang, and Luo identified many of these positive trends. But one negative aspect they identified, which is still an issue today, is "the loss of fossil resources. . . . Spontaneous excavation of fossil resources by local citizens for economic gain is not an uncommon phenomenon in China . . . many fossils were collected by a growing number of private collectors and have become inaccessible for scientific research. It is a great challenge for the government at all levels and for the research community." Mining activity has also become an issue. A 2017 article in *Nature* described how the Doushantuo site is being endangered by phosphate mining. But the current focus on the sale of fossils is on amber from China's neighbor, Myanmar.

Amber is one of the wonders of the world. As the plant resin that forms amber hardens, it can trap organisms living around it and preservation can be exquisite. There are about two hundred amber deposits worldwide, with the best known being from the Baltic region and the Dominican Republic.[20] In the last decade, attention to amber from Myanmar, called Burmese amber, has exploded. Dating to the Cretaceous, these deposits had yielded 1,200 species by 2018, including representatives of 569 families of insects.[21] As remarkable as the insect species are, including the huge-jawed "hell ants," it is the rare vertebrate specimens that have gotten the most attention. These include baby frogs and snakes, the feathered tail of a baby dinosaur, bird feet, and the exquisite skull of a tiny lizard (originally thought to be a bird). Receiving as much attention as the fossils themselves, however, is the way they are collected and sold.

Most of the amber comes from Kachin State in Myanmar, where it is mined by the locals, often in unsafe conditions.[22] The specimens are then smuggled across the border into China and sold at a market in Tengchong. Rare specimens can be worth tens of thousands of dollars and up. Although most specimens are of relatively little commercial value, there are so many of them that the total trade may be up to $1 billion. Given their scientific value, many paleontologists, including some in China, have held their noses and paid for the specimens at increasingly inflated prices. But that is not the only issue.

Until 2017, much of the income from the amber mines benefited a local ethnic insurgency group. In June 2017, the army of Myanmar seized control. Given the army's reputation for human rights abuses elsewhere, an immediate shadow was thrown over the sale of amber, with the understandable fear that it would fund the army's abuses.[23] With the best intentions, the Society for Vertebrate Paleontology expressed its deep concerns and in April 2020 sent a letter to the editors of paleontology journals requesting "a moratorium on publication for any fossil specimens purchased from sources in Myanmar after June 2017 when the Myanmar military began its campaign to seize control of the amber mining."[24] A 2020 paper on a fossil amphibian specifically stated that "specimens were acquired following the ethical guidelines for the use of Burmese amber set forth by the Society for Vertebrate Paleontology. . . . We hope that this study will serve as a model for other researchers working with these types of materials in this region."[25]

Nonvertebrate paleontologists, especially those who work on arthropods, were not so agreeable to the SVP letter. While acknowledging the deeply disturbing humanitarian situation in Kachin State, fifty-five paleontologists from twenty countries argued in a sharply worded statement that "the situation surrounding Burmese amber is not as simple as the SVP letter suggests. A general moratorium on fossils from this type of amber is not justified by the facts presented . . . and is unlikely to improve the situation in Myanmar. It will instead have a very negative impact on practical research, scientists, the institutions that support them, and even the community in Myanmar that makes a living from harvesting these ambers."[26] They also expressed concern that the SVP "is to a certain extent geographically and systematically restricted, and it does not speak for the entire palaeontological community." A parallel statement by the International Palaeoentomological Society (fossil insect workers) also opposed the SVP call for a ban but laid out possible steps to address the problem.[27] I hope someone is organizing a joint meeting of all stakeholders to move this situation forward in a positive way.

In Latin America, significant paleontological research is being carried out in Brazil, Peru, Mexico, and Bolivia. The rich fossil resources of Argentina became known to the English-speaking world through Darwin's visits in the 1830s, described in *The Voyage of the Beagle* (1839), and George Gaylord Simpson's expeditions in the 1920s and 1930s, colorfully described in *Attending Marvels* (1934). As discussed by Diana Elizabeth Fernández, Leticia Luci, Cecilia Soledad Cataldo, and Damián Eduardo Pérez in 2014 and updated with me in 2021, paleontological research in Argentina has a deep history of research by both native and foreign scholars.[28] The key South American paleontological journal, *Ameghiniana*, is named after one of them—Florentino Ameghino, considered the father of Argentinian paleontology. It is published by the Asociación Paleontológica Argentina, the professional society, which has twenty-three local chapters. Most Argentinian paleontologists work as researchers with CONICET (Consejo Nacional de Investigaciones Científicas y Técnicas), and the science was reasonably well funded until 2016. That year saw a major crash (about 35 percent) in government funding for science and technology, which has persisted. The value of existing research project funds has been eroded by high annual inflation rates. CONICET researchers saw their incomes drop by 45 percent between 2015 and 2020.

Three universities support degree work in paleontology: the University of Buenos Aires (UBA), the National University of Río Negro, and the National Comechingones University; all attract students from elsewhere in Latin America. In 2019 UBA began to offer a PhD in paleontology. The National University of Chubut has begun offering a three-year paleontology technician degree, focusing on preparation and display; I believe this is unique in the world. All of these degrees are tuition-free.

There are significant fossil collections throughout Argentina, but many of the most important fossils are in foreign museums. The country has become much more proactive in response to the illegal trafficking of fossils. As Fernández and her colleagues point out, "countless fossils have been illegally taken from our country in the last 100 years. Under the updated, current law for protection of paleontological heritage, progress on the subject has been made, but a lot remains to be achieved."

Brazil, like Argentina, has issues with the illegal trade in fossils despite tough laws against it.[29] In 2019 a French judge ordered the return to Brazil of forty-five fossils, including a nearly complete pterosaur skeleton from the Araripe Basin in northeast Brazil, the source of many spectacular Cretaceous-age fossils.[30] In late 2020, another contretemps broke out over the feathered dinosaur *Ubirajara jubatus* from the same area, which was transported to Germany and housed in a museum there. Brazilian paleontologists have raised doubts that proper permissions were obtained and have pushed for repatriation.[31] These issues are on top of the devastating 2018 fire that destroyed much of the Museu Nacional in Rio de Janeiro, ruining the paleontology collection and destroying irreplaceable cultural artifacts.

Although the most publicized fossils from Africa are those of early humans, rich troves of fossils representing the entire history of the Earth come from the continent. Probably the most familiar of those are the seemingly countless fossils from the Paleozoic of southern Morocco, especially stunning Devonian-age trilobites and nautiloid cephalopods, which I have seen in every rock shop I have visited. It is estimated that the total trade is worth about $40 million a year and supports more than fifty thousand people.[32] Not all of these fossils are real; one article in the *New York Times* suggested that up to 20 percent were fakes (some of the ones I have seen are quite good).[33] Morocco has also yielded one of the most important new fossil localities found in recent decades, the extraordinary Ordovician Fezouata biota and its spectacular soft-bodied fossils.

Egypt has also yielded numerous fossil treasures, especially fossil vertebrates, including dinosaurs. Like many other countries, until recently fossil research was conducted by foreigners. Hesham Sallam founded the Mansoura University Paleontology Center (MUVP), the first vertebrate paleontology program in the Middle East in 2010, and in 2018 he described the first Late Cretaceous sauropod from North Africa. Sallam is notable for his willingness to train woman as paleontologists; in 2019 I met his student, the remarkable Sanaa El-Sayed, the first female in her country to receive a graduate degree in paleontology.

At the other end of the continent, South Africa also has a long history of paleontological research. South African fossils, such as the Permian plant *Glossopteris*, were key pieces to solving the puzzle of plate tectonics. Vertebrate and plant fossils of the Karoo Basin are critical for interpreting changes in terrestrial environments across the Late Paleozoic to Early Mesozoic, especially the end-Permian mass extinctions. Some of the oldest life on Earth comes from the 3.26-billion-year-old Fig Tree Group. The Nama

Basin in neighboring Namibia provides important evidence for changes at the Ediacaran-Cambrian transition and thus the origin of animals. And again, there are renowned sites for early hominids (and yes, dinosaurs). In 2013 the University of the Witwatersrand launched the South African Strategy for the Palaeosciences and awarded the Centre of Excellence for the Palaeosciences (CoE-Palaeo) to the university.[34] The strategy document is frank in addressing the impact of apartheid on archaeology and paleontology. CoE-Palaeo is "a global hub for the study of the origins of species, using cutting-edge research techniques to understand South Africa's unique fossil and archaeological record."[35] As best I can tell, the organization exists mainly to support graduate students and postdocs. I recently heard an excellent talk on *Glossopteris* by Aviwe Matiwane, a Black woman PhD at Rhodes University who has been funded by CoE-Palaeo.[36] Matiwane has a strong social media presence as a science communicator.

Australia also has a small but vital paleontology community. As is the case in the UK, a number of Australian-born paleontologists have established themselves in the United States and become leaders in the field here, and some American paleontologists have reversed the long journey. The regional paleontological organization, the Australasian Palaeontologists, publishes the journal *Alcheringa*, which attracts contributions from well beyond Australia. The largest paleontology program in the country is the Palaeoscience Research Centre at the University of New England. In 1942, a remarkable group of soft-bodied fossils were discovered in the Ediacara Hills of the Flinders Ranges in South Australia. These fossils date to about 560 million years ago, well before the beginning of the Cambrian. Also, in the Flinders Ranges is a site where a thick carbonate unit can be found directly above the glacial deposits characteristic of Snowball Earth, dated at about 635 million years ago. In 2004, the period between 635 and 541 million years ago was officially designated as the Ediacaran Period, after the site where the fossils were found. The Ediacaran fossils in Australia are an active place of research for researchers from Australia and around the world. The original site is now protected as the Ediacara Conservation Park. At the other end of the geological time scale, recent years have revealed new dinosaurs, including some spectacular tracksites.

Even more remote from Australia is New Zealand, which has a richer fossil record than one would expect for a small country, especially for fossils from the Cenozoic. James Crampton and Roger Cooper reviewed the status of paleontology in the country in 2010 and counted thirty-three paleontologists there.[37] About half of those worked for government agencies, with the rest scattered around various universities. Since then Crampton told me that this number has markedly shrunk[38]. Their work is supplemented by the efforts of committed amateurs. One key achievement is the Fossil Record File, which in 2020 contained detailed information on more than one hundred thousand fossil localities in New Zealand, the South Pacific islands, and parts of Antarctica.[39] This remarkable database is an essential resource for understanding the evolution of life in New Zealand and the Southwest Pacific, and it has been used worldwide, for example, to quantify the completeness of the fossil record. Funding has been static for decades, with most funded research exploring biostratigraphy (fossil fuels) and the paleoclimate, i.e., "applied paleontology." A major impact regarding fossil fuels was an April 2018 halt in new permits for offshore oil exploration. Unfortunately, like so many other places in the world, there has been an almost total loss of taxonomic expertise. According to Crampton, "there is now no one in New Zealand actively researching any fossil group older than Late Cretaceous, and to all intents, very little significant work is going on in anything older than the Miocene. There is very little alpha taxonomic work being undertaken at all—most paleontology is using existing taxonomies and/or data sets to explore issues of environmental change, and most relates to the Quaternary. There is a paleobotanist employed at GNS,[40] and I could nominally be called a mollusc specialist although I never work in this area now, and there are a couple of other retired or otherwise employed macropaleontologists, but apart from these people there is no research being undertaken on anything other than microfossils."

The nature of international science has changed dramatically in the decades since my first conference in 1983. For better or worse, English has established itself as the international language of science. It has been many years since I have had to puzzle out a paper written in German, French, or worst of all, Russian (not the same alphabet!). Nearly every scientific paper, no matter where published, is now written in English or at least has an English translation of the abstract. I also rarely must track down the paper copy of an article published in a foreign journal. Most are now available online through the university library, clearinghouses such as Research Gate, or can be requested from helpful colleagues. The institution of the paper reprint, which was requested with a postcard and had to be mailed, is effectively extinct.

Using letters for communication is also essentially extinct. I have kept some prized letters from colleagues in other countries, received at a time when getting an airmail letter from abroad was an event (and you saved the stamps). Now I can send a quick email anywhere in the world.

International conferences are also largely given in English, even those that take place in countries where that is not the native language. This is clearly an imposition on many of my foreign colleagues, who sometimes struggle to present a talk in a strange tongue, especially one with as many nonintuitive aspects as English. The slides need to tell the story.

One large conference is the North American Paleontological Convention (NAPC). The most recent, held in 2019 in Riverside, California, was attended by nearly seven hundred paleontologists from thirty-three countries around the globe. Of the 178 who came from outside the United States, the largest delegations were from China (31), Canada (25), the UK (20), and Germany (16).[41]

Unlike the large general meetings at the NAPC, most international conferences are focused on a single area of research interest. In recent years I have participated in conferences on trace fossils, unusual preservation, and the Ordovician. Because these are small and focused, there is a much greater opportunity to get to know your colleagues well and to develop both research collaborations and personal friendships. Local field trips are a good way to learn about some new rocks and fossils. And these meetings are a great excuse to travel; I often take vacation time right after a conference is over.

Unfortunately, international meetings are also subject to political issues that have nothing to do with the science. Fred Ziegler at the University of Chicago loved to tell us about the Silurian meeting he attended in Prague in 1968, just as the Soviet tanks rolled into Czechoslovakia. Today there are major impediments to international travel, and foreign scholars may find it difficult or impossible to get visas to visit the site of a conference.

There are also the issues of cost and travel time. Not long ago I really wanted to attend a conference in Patagonia, Argentina. Then I learned it would take about thirty hours of total travel time to get there and cost upward of $3,000; I decided to take a pass. The issue of cost is certainly far worse for scholars in poorer countries. At a meeting I attended in Newfoundland, many of the talks by scientists from India and China were canceled after they were scheduled because the speakers decided not to come. I suspect time and money were the key culprits, but it left the organizers scrambling to fill the empty slots in the schedule.

For these reasons, it is probable that many of the changes to meetings forced by the ongoing pandemic may become permanent (see chapter 14). In September 2020, I attended a four-day conference on the Ordovician, all virtual. Each of the sessions started at a different time, from 1 a.m. to 9 a.m. (Central time), to accommodate the schedules of scientists across the globe. Those of us who did not want to get up at four o'clock in the morning were thankful that the sessions were recorded. Overall, it worked quite well. Being both free and quite valuable, with no travel costs to the participants, encouraged true international participation.

Similarly, a group of early career scientists have organized a weekly free online seminar series called Pal(a)eo PERCS for speakers who are just starting out.[42] This has provided an unrivaled introduction to the accomplishments of young scientists across the spectrum of paleontology and related sciences. Live streamed short seminars are followed by networking opportunities through breakout rooms and "teatime" with the speaker. Because it is online, there have been presenters from the UK, Sweden, Norway, South Africa, Peru, and India, as well as the United States, all without anyone having to leave the security of home. One nice feature is a clear Code of Conduct that governs online interactions: they "value the presentations and attendance of diverse scientists from a range of backgrounds and experiences. As a result, this seminar is committed to maintaining a harassment-free environment for everyone. Accordingly, participants are expected to treat one another with respect and dignity." I have learned a lot from these presentations, have enjoyed the virtual interactions, and am excited to meet this wonderful and diverse group of scientists, the rising generation of researchers. Our discipline is in safe hands!

22

LAST THOUGHTS

SWOT-ing at Paleontology

I t's hard not to preen a little when someone asks what I do
and I reply, "I'm a paleontologist." Paleontology is one of the
most popular topics in science in the public and the media.
Just this morning BBC radio had a story about dinosaurs. A quick
check of the *New York Times* website reveals nearly four thousand articles under the keyword "paleontology"; a similar search
on LiveScience.com yields thousands of hits on the word "fossils." New paleontological discoveries are prominently featured
on the homepage of the National Science Foundation. We paleontologists clearly have numerous opportunities to interest and
excite the public about our research. It may seem contradictory,
therefore, that the perception shared by many paleontologists is
that our field is threatened by an existential crisis, that funding
for paleontological research has become effectively nonexistent,
and that hiring young paleontologists to replace those who have
retired has dropped below sustainable levels.

Those with experience preparing strategic plans for busi-
nesses and organizations may have noticed my deliberate use of
the terms "opportunities" and "threat." These are usually coupled
with "strengths" and "weaknesses" to form the acronym SWOT
(Strengths, Weaknesses, Opportunities, Threats), which describes

an approach to identifying the factors that have the most impact on an entity. Strengths and weaknesses generally refer to internal factors, whereas opportunities and threats comprise external influences.

In this book, I have covered many of the strengths, weaknesses, opportunities, and threats to paleontology as I and my colleagues see them. In this final brief chapter, I attempt to summarize them into an admittedly incomplete SWOT analysis on paleontology as a discipline and focus on the structural aspects of the field rather than on new research directions.

STRENGTHS

The most obvious strength of paleontology is its continued intellectual productivity. At every Geological Society of America meeting, there may three or even four paleontology-themed sessions going on simultaneously. Hundreds of other presentations are given at the Society of Vertebrate Paleontology annual meetings. Nearly seven hundred paleontologists from around the world attended and presented papers at the 2019 North American Paleontological Convention.

Paleontology journals are similarly flooded with submissions and publish hundreds of papers each year. Paleontology papers also frequently appear in high-profile journals such as *Science, Nature, Geology*, and the *Proceedings of the National Academy of Sciences*. This is a testament both to the continued pace of new and significant discoveries and to major advances in analyzing the known fossil record, such as major extinction events.

Student interest also remains high, with many talented young people entering the field every year. The professional societies have made major efforts to support and reward their investigations. I have been strongly impressed by the incredibly bright and productive young people I have encountered and heard present their research.

The flood of new discoveries continues, from across the world. Many of these are from China, which has grown to be a major player in paleontology on the global stage.

Paleontology has become increasingly integrated with the biological sciences. The growth of conservation paleobiology has highlighted the science's relevance to modern global problems. And data from the fossil record remains critical for understanding past and future global climate change and its impact on the biosphere.

Paleontologists have also been active in the development of new databases and online resources. The Paleobiology Database makes the "raw stuff" of the fossil record and tools for analyzing it accessible to a wide audience. We have also been open to and eager to use new technologies as they become available.

Outreach to educators is another important strength. Educational materials about the field are readily obtained through sites run by the University of California Museum of Paleontology, and the Paleontological Research Institution and the Museum of the Earth both have an active and ongoing educational component. We have also become more welcoming to nonprofessional fossil enthusiasts.

WEAKNESSES

A critical global problem for the field is the erosion of expertise in the general area of descriptive and systematic paleontology. Although some groups have multiple specialists working on them, others are effectively down to their last expert with no prospects of replacements being trained. A side effect of this decline is the tension between the remaining "classically" trained paleontologists and those whose interests are more analytical or theoretical. This is unfortunate because the two approaches can and should be mutually illuminating. Without taxonomic expertise, it is not possible to assess the quality of data entered into any current or future database or used in synoptic analyses of evolution in the fossil record. At the same time, analytical approaches can highlight those areas of the fossil record that need detailed and informed taxonomic work.

A related weakness is the marked decline of paleontology at major oil companies. At one time, most companies had large in-house paleontology programs for paleoecology and biostratigraphy, including associated databases and collections. Today many of these programs are either gone or much smaller, and much of the work is contracted out. Those jobs are gone.

Another concern is the state of paleontology's professional societies. Traditionally, a major role of the societies was to promote the field by organizing meetings and publishing journals. For many of us, the major incentive to join a society was to receive a personal copy of its journals. Over the last two decades, however, nearly all of the journals have become available electronically through college and university libraries. Probably as a result, many scientific societies, not just those in paleontology, have seen decreases in membership. A critical task for the societies will be to create new incentives for being a member.

Even more critical is the lack of internal unity in the field and a concomitant lack of a common voice. A paleontologist might belong to the Paleontological Society, the Society for Vertebrate Paleontology, the Paleobotanical Section of the Botanical Society, or the Society for Sedimentary Geology (SEPM), but rarely to more than one of these. These societies seldom meet together and have relatively little formal contact. Paleontologists with interests in paleoclimatology or paleoceanography may attend the American Geophysical Union meeting rather than the Geological Society of America. Except for the infrequent North American Paleontological Conventions, paleontologists rarely, if ever, meet as a unified group.

This disunity within the field has made long-term planning and follow-through on research initiatives difficult, and efforts to generate support for these initiatives through the National Science Foundation, the wider scientific community, the public, or members of Congress has been disappointing. Efforts to create unified long-term plans have all sputtered out.

Professional paleontologists have not yet come up with a unified and nuanced approach to the issues posed by commercial fossil collecting, fossil sales, and access to public lands. Again, we need to speak with a single voice to the numerous stakeholders.

The last decade has thrown light on a long-hidden weakness, our lack of diversity. Although there has been a notable improvement in gender diversity, we remain one of the least racially and ethnically diverse areas of science. Our professional societies and funders are just now beginning to deal with this issue.

OPPORTUNITIES

Paleontology remains one of the most popular fields of science among the general public; after all, who doesn't love dinosaurs? The construction of the

Museum of the Earth and of the new paleontology exhibits at places such as the Smithsonian are testimony to this abiding interest. We need to further capitalize on this interest.

We also have an active and enthusiastic amateur community. Paleontology, along with astronomy, remains one of the few fields in which the trained amateur can make a fundamental discovery and be recognized for it by the professional community. We need to extend our efforts to integrate avocational paleontologists into the discipline and encourage them to be advocates for our science.

The importance of paleontology is now recognized by many in the broader scientific community. Paleontologists have moved into new areas of expertise as our "parent" fields of geology and biology have changed over time. Paleontological expertise is an essential part of studies of the dynamics and history of the entire Earth system, and fruitful collaborations have been established with geoscientists in areas such as paleoclimatology and biogeochemistry. Similarly, the paleontological record of early life is an integral part of areas such as evolutionary developmental biology (evo-devo) and studies of the phylogeny of major groups. In the latter case, paleontological data has played a critical role in constraining the timing of the origin of new taxa based on molecular data. These interactions need to be deepened and broadened. We need to make our colleagues our advocates.

THREATS

Paleontology finds itself threatened on many fronts. First, our central tenet, that life has a long and complicated history, is taught in a country where a large fraction of the adult population does not accept the validity of evolution. A large proportion also does not accept the reality or the seriousness of global climate change.

The lack of funding for and positions in paleontology are continuing deep concerns. The number of potential financial supporters for paleontological research is quite limited. Private and foundation funding is virtually nonexistent, and federal funding has long been static. As a result, paleontologists have not been replaced as they retired, with preference going to fields with greater perceived funding possibilities. There is also the perception—which I hope this book has helped to dispel—that we are an old-fashioned and outmoded science.

Karl Flessa and Dena Smith compared changes in academic employment between 1980 and 1995 among paleontologists, based on data in the American Geological Institute (AGI) *Directory of Geoscience Departments*.[1] They noted that the typical paleontologist was becoming older, as the number of untenured (assistant professor) faculty decreased.

With the help of Cindy Martinez of AGI, in 2007 I compiled the distribution of ranks among the three disciplines in the current version of the directory. In that year, there were 81 assistant professors and four-times as may full professors (333), as well as 184 emeritus faculty. In comparison, among geochemists there were 129 assistant professors and 313 full professors and 100 emeritus faculty. The field of paleontology is definitely graying!

TOWARD THE FUTURE

A SWOT analysis is only a place to start; it does not address the key question of what is to be done. To me, the issue is not whether we can identify important and interesting areas of research; we have done this repeatedly (see chapter 15). Instead, both individuals and our institutions need to look at long-term structural changes that will make them more effective advocates for our science. We need to find a way to speak with a unified voice on the issues that matter to us. This will require much closer cooperation among our disparate societies. We need to present a united front on issues such as commercial collecting, access to land, diversity, and funding. Advocacy is far more effective when there is a single clear message.

The paleontological community should take the steps necessary to develop a true strategic plan, one that can guide the field over the next decade and beyond. The astronomy community already does this on a routine basis. Key elements of this plan should include an expanded form of this SWOT analysis to establish the current context of the field; a statement of overarching goals that identifies the major scientific problems the field should be focused on in the next ten years; and an assessment of the infrastructure and expertise needed to attain these goals. We can then tell the funding agencies, "this is what the community wants and why you should support it." Some of this has been done before, but what has been lacking is the follow-through necessary to propel the plan beyond any initial problems. Again, this will require our professional societies to commit to working together for the long term.

We also need more individuals who can effectively represent the science of paleontology to scientists outside our community and to the public. We need to change the perception of paleontologists from white men with beards digging up dinosaurs to reflect the true diversity of our community and their scientific interests. Paleontologists need to be positive, not defensive, about the fossil record and what it tells us. We need to emphasize that paleontology not only studies the life of the past but has critical insights into life today and into the future. We need to convince deans and department heads that any biology or geology department that lacks a paleontologist damages their scientific mission and restricts the training of their students. We may study the past, but we are a young, vibrant, and essential field with much to contribute to the future!

ACKNOWLEDGMENTS

The bulk of this book was written while I was an Edward P. Bass Distinguished Environmental Visiting Scholar at Yale University; their support is gratefully acknowledged. I am deeply indebted to the Yale Department of Earth and Planetary Sciences and the Yale Institute for Biospheric Studies (YIBS) for hosting me, and to Derek Briggs, who helped arrange my visit. He and his wonderful spouse Jennifer Briggs also went out of their way to make me feel at home. Appreciation also goes to the members of the Yale Department of Earth and Planetary Sciences and the Yale Peabody Museum faculty and staff who made time to talk with me and put up with my nearly constant presence: Pincelli Hull, Bhart-Anjan Bhullar, Jacques Gauthier, Lidya Tarhan, Ellen Thomas, Susan Butts, Jessica Utrup, Christina Lutz, Marilyn Fox, Eric Lazo-Wasem, and Lourdes Rojas. Many present and former Yale graduate students and postdocs willingly gave up some of their precious time to chat with me: Ross Anderson, Gwen Antell, Janet Burke, Robin Canavan, Elizabeth Clark, Leanne Elder, Michael Henehan, Nicholas Mongiardina Koch, Holger Petermann, Sarah Tweedt, Jasmina Wiemann, Sophie Westacott, and Christopher Whelan.

Colleagues and students at other institutions were equally generous with their time, often giving far more fulsome answers than I had a right to expect or could possibly include (and correcting errors on my part). I cannot thank them enough (and I hope I left no one out): John Alroy,

Jennifer Bauer, Anna Kay Behrensmeyer, Carl Brett, David Bottjer, Lisa Park Boush, Lawrence Bradley, Julia Brunner, Kate Bulinski, Patricia (P. J.) Burke, Sandra Carlson, Cecilia Soledad Cataldo, Phoebe Cohen, James Crampton, Ellen Currano, Benjamin Dattilo, Stephen Dornbos, Matthew Downen, Mary Droser, Lucy Edwards, Sara ElShafie, Diana Elizabeth Fernández, Karl Flessa, Margaret Fraiser, Rebecca Freeman, Robert Gastaldo, Patrick Getty, Alan Goldstein, Emily Graslie, Howard Harper, David Jablonski, Patricia Kelley, Susan Kidwell, Wolfgang Kiessling, Paul Koch, Michał Kowalewski, Jim Lehane, Scott Lidgard, Jere Lipps, Rowan Lockwood, Leticia Luci, Zhe-Xi Luo, Bruce MacFadden, Jacalyn Wittmer Malinowski, Pedro Marenco, Charles Marshall, Herb Meyer, Greg McDonald, Arnold Miller, Philip Novack-Gottshall, Judith Parrish, Marty Perlmutter, Damián Eduardo Pérez, P. David Polly, Sara Pruss, Vincent Santucci, Sarah Sheffield, Thomas Servais, Sally Shelton, Dena Smith, Steve Stanley, Alycia Stigall, Jessica Theodor, Marc Uhen, Sally Walker, Peter Wagner, Johnny Waters, Anne Weil, Lisa White, Scott Williams, and Shuhai Xiao.

A coterie of talented paleoartists generously spent time discussing their work with me: Jimi Catanzaro, Marianne Collins, Wendy Luck, Darren Naish, Barbara Page, Gary Raham, Rachel Rasfeld, Luis Rey, Evan Sizemore, Gary Staab, William Stout, and Ray Troll.

I deeply appreciate the cooperation of amateur paleontologists and fellow ESCONI members John Catalani, Jack Wittry, and David Carlson, as well as the talented and enthusiastic Riley Black and Joseph (PaleoJoe) Kchodl. Thanks also to Linda McCall and Kurt Zahnle. Bruce and Rene Lauer have been gracious with their time.

Jennifer Verdolin and Sandra Novack-Gottshall were my invaluable guides to the perplexing world of book publishing. I am grateful to my editor at Columbia University Press, Miranda Martin, for taking a chance on a first-time author who was not a celebrity. The comments of Victoria Arbour, Lindsay Zanno, and an anonymous reviewer greatly improved the manuscript. Kathryn Jorge, production editor at Columbia University Press, and Ben Kolstad, project manager at KnowledgeWorks Global Ltd., are thanked for turning my flawed manuscript into the book you are reading.

My sons, Daniel and Jonathan, are a constant source of pride and have been supportive every step of the way. Words cannot express the gratitude I owe to my wife and best friend Deb Stewart, who has lovingly encouraged me throughout this project and my entire career and has had almost infinite patience with my idiosyncrasies.

NOTES

1. THOSE WHO KNOW THE PAST

1. J. McPhee, *Basin and Range* (New York: Farrar, Straus and Giroux, 1982). A must-read for anyone interested in how geology works.

2. WE HAVE THE BEST QUESTIONS

1. Stephen Jay Gould's *Wonderful Life* (New York: Norton, 1989), although badly outdated, is still the most compelling introduction to the Burgess Shale (and to the fertile mind of Steve Gould).
2. S. J. Gould, "Eternal Metaphors of Palaeontology," *Developments in Palaeontology and Stratigraphy* 5 (1977): 1–26.
3. Entire books have been written about punctuated equilibria, its scientific support, and its reception. It remains controversial to this day. See N. Eldredge, *Time Frames: The Re-Thinking of Darwinian Evolution and the Theory of Punctuated Equilibria* (New York: Simon & Schuster, 1985). Also see the original paper: N. Eldredge and S. J. Gould, "Punctuated Equilibria: An Alternative to Phyletic Gradualism," in *Models in Paleobiology*, ed. T. J. M. Schopf (San Francisco: Freeman, Cooper, 1972), 82–115.
4. R. M. Hazen, D. Papineau, W. Bleeker, R. T. Downs, J. M. Ferry, T. J. McCoy, D. A. Sverjensky, and H. Yang, "Mineral Evolution," *American Mineralogist* 93, nos. 11–12 (2008): 1693–1720.
5. S. Conway Morris, "Evolution: Like Any Other Science It Is Predictable," *Philosophical Transactions of the Royal Society B: Biological Sciences* 365, no. 1537 (2010): 133–45.

6.   N. Shubin, *Your Inner Fish: A Journey Into the 3.5-Billion-Year History of the Human Body* (New York: Vintage, 2009).

3. I'M NOT ROSS (OR INDIANA JONES)

1.   For a detailed discussion of the training and responsibilities of mitigation paleontologists, see P. C. Murphey, G. E. Knauss, L. H. Fisk, T. A. Deméré, and R. E. Reynolds, "Best Practices in Mitigation Paleontology," *Proceedings of the San Diego Society of Natural History* 47 (2019): 1–43.
2.   M. B. Farley and J. M. Armentrout, "Fossils in the Oil Patch," *Geotimes* 45, no. 10 (2000):14.
3.   R. E. Plotnick, "A Somewhat Fuzzy Snapshot of Employment in Paleontology in the United States," *Palaeontologia Electronica* 11, no. 1 (2008).

4. ATTENDING MARVELS

1.   I got a lot of papers out of this! Here are some: G. D. Cody, N. S. Gupta, D. E. G. Briggs, A. L. D. Kilcoyne, R. E. Summons, F. Kenig, R. E. Plotnick, and A. C. Scott, "Molecular Signature of Chitin-Protein Complex in Paleozoic Arthropods," *Geology* 39, no. 3 (2011): 255–58; W. T. Fraser, A. C. Scott, A. E. S. Forbes, I. J. Glasspool, R. E. Plotnick, F. Kenig, and B. H. Lomax, "Evolutionary Stasis of Sporopollenin Biochemistry Revealed by Unaltered Pennsylvanian Spores," *New Phytologist* 196, no. 2 (2012): 397–401; R. E. Plotnick, F. Kenig, A. C. Scott, I. J. Glasspool, C. F. Eble, and W. J. Lang, "Pennsylvanian Paleokarst and Cave Fills from Northern Illinois, USA: A Window Into Late Carboniferous Environments and Landscapes," *PALAIOS* 24, no. 9–10 (2009): 627–37; A. C. Scott, F. Kenig, R. E. Plotnick, I. J. Glasspool, W. G. Chaloner, and C. F. Eble, "Evidence of Multiple Late Bashkirian to Early Moscovian (Pennsylvanian) Fire Events Preserved in Contemporaneous Cave Fills," *Palaeogeography Palaeoclimatology Palaeoecology* 291, no. 1–2 (2010): 72–84.
2.   ACME mapper: https://mapper.acme.com/.
3.   TopoView: https://ngmdb.usgs.gov/topoview/.
4.   For two of their many papers on the subject, see M. Kowalewski, G. A. Goodfriend, and K. W. Flessa, "High-Resolution Estimates of Temporal Mixing Within Shell Beds: The Evils and Virtues of Time-Averaging," *Paleobiology* 24, no. 3 (1998): 287–304; S. M. Kidwell, "Time-Averaged Molluscan Death Assemblages: Palimpsests of Richness, Snapshots of Abundance," *Geology* 30, no. 9 (2002): 803–6.
5.   Here is a sampling of Patricia Kelley's papers on predator-prey interactions: P. H. Kelley and T. A. Hansen, "Naticid Gastropod Prey Selectivity Through Time and the Hypothesis of Escalation," *PALAIOS* 11, no. 5 (1996): 437–45; A. A. Klompmaker, P. H. Kelley, D. Chattopadhyay, J. C. Clements, J. W. Huntley, and M. Kowalewski, "Predation in the Marine Fossil Record: Studies, Data, Recognition, Environmental Factors, and Behavior," *Earth-Science Reviews* 194 (2019): 472–520; C. C. Visaggi and P. H.

Kelley, "Equatorward Increase in Naticid Gastropod Drilling Predation on Infaunal Bivalves from Brazil with Paleontological Implications," *Palaeogeography Palaeoclimatology Palaeoecology* 438 (2015): 285–99.

6. Martin Brasier, *Darwin's Lost World: The Hidden History of Animal Life.* (Oxford: Oxford University Press, 2010).

7. B. F. Dattilo, D. L. Meyer, K. Dewing, and M. R. Gaynor, "Escape Traces Associated with *Rafinesquina alternata*, an Upper Ordovician Strophomenid Brachiopod from the Cincinnati Arch Region," *PALAIOS* 24, no. 9–10 (2009): 578–90.

8. As cited in Greg Hand, "Moore Medal Honors UC's Brett," University of Cincinnati, April 30, 2012, https://www.uc.edu/profiles/profile.asp?id=15719.

9. P. D. Ward, *On Methuselah's Trail: Living Fossils and the Great Extinctions* (New York: W. H. Freeman, 1991).

10. One member of the party apparently packed a lawn chair rather than his tent!

11. G. G. Simpson, *Attending Marvels: A Patagonian Journey* (New York: Time, 1965), 165.

12. But the trips are worth it. See S. Q. Dornbos, T. Oji, A. Kanayama, and S. Gonchigdorj, "A New Burgess Shale-Type Deposit from the Ediacaran of Western Mongolia," *Scientific Reports* 6, no. 23438 (2016), https://doi.org/10.1038/srep23438; D. R. Cordie, S. Q. Dornbos, P. J. Marenco, T. Oji, and S. Gonchigdorj, "Depauperate Skeletonized Reef-Dwelling Fauna of the Early Cambrian: Insights from Archaeocyathan Reef Ecosystems of Western Mongolia," *Palaeogeography Palaeoclimatology Palaeoecology* 514 (2019): 206–21.

13. Snowball Earth, more formally known as the Cryogenian, is a period when, at least twice, glaciers reached as far as the equator and the entire planet was more or less frozen. For more information, see SnowballEarth.org at http://www.snowballearth.org/.

14. D. Paquette, "The Pandemic Has Left a Huge Cache of Dinosaur Bones Stuck in the Sahara," *Washington Post*, February 12, 2021, https://www.washingtonpost.com/world/2021/02/12/niger-dinosaur-bones/.

15. K. B. H. Clancy, R. G. Nelson, J. N. Rutherford, and K. Hinde, "Survey of Academic Field Experiences (SAFE): Trainees Report Harassment and Assault," *PLoS One* 9, no. 7 (2014): e102172.

16. Paleontological Society, "Non-Discrimination and Code of Conduct," ratified April 29, 2019, https://www.paleosoc.org/paleontological-society-policy-on-non-discrimination-and-code-of-conduct/.

17. But Arnold Miller did get a nice paper out of the experience. See A. I. Miller, G. Llewellyn, K. M. Parsons, H. Cummins, M. R. Boardman, B. J. Greenstein, and D. K. Jacobs, "Effect of Hurricane Hugo on Molluscan Skeletal Distributions, Salt River Bay, St. Croix, U.S. Virgin Islands," *Geology* 20, no. 1 (1992): 23–26.

18. This is one of the most highly cited papers in paleontology (more than 1,250 times as of August 2019): A. K. Behrensmeyer, "Taphonomic and Ecologic Information from Bone Weathering," *Paleobiology* 4, no. 2 (1978): 150–62. Forty years later, she is still producing highly significant work: A. K. Behrensmeyer, "Four Million Years of African Herbivory," *Proceedings of the National Academy of Sciences* 112, no. 37 (2015): 11428–29.

19. For a description of the project after more than a decade, see E. N. Powell, G. M. Staff, W. R. Callender, K. A. Ashton-Alcox, C. E. Brett, K. M. Parsons-Hubbard, S. E. Walker, and A. Raymond, "Taphonomic Degradation of Molluscan Remains During Thirteen

Years on the Continental Shelf and Slope of the Northwestern Gulf of Mexico," *Palae-ogeography Palaeoclimatology Palaeoecology* 312, no. 3–4 (2011): 209–32.

5. SAFE PLACES

1. The holotype is a single specimen to which the name of a species is attached; it physically represents the author's concept of a specimen that best represents the species. In this case, it is Yale Peabody Museum specimen 1980, which is its number in the museum catalog. *Brontosaurus excelsus* is the species name, given by O. C. Marsh in 1879. Holotypes are irreplaceable and are closely guarded.
2. For a profile of Paul Mayer, see https://www.fieldmuseum.org/about/staff/profile/91.
3. Society for the Preservation of Natural History Collections, https://spnhc.org/.
4. Association for Materials and Methods in Paleontology, https://paleomethods.org/.
5. Writing in the *Chronicle of Higher Education* in 2006, Thomas Benton bemoaned the update of the Dinosaur Hall at Philadelphia's Academy of Natural Sciences: "The towering *Hadrosaurus* is . . . banished to an inconspicuous corner to make room for a gathering of fossil replicas designed as photo-ops. Instead of gazing up at a relic of the heroic era of Victorian science, people ignore the *Hadrosaurus* and get their picture taken with their head beneath the jaws of the scary *Giganotosaurus* . . . before going to the gift shop to buy a 'sharp toothed' plush toy. See, kids, science can be fun!" Thomas H. Benton, "The Decline of the Natural-History Museum," *Chronicle of Higher Education*, October 13, 2006.
6. "Understanding Global Change," University of California, Berkeley, https://ugc.berkeley.edu/
7. The human tendency to see familiar things where they do not exist is known as pareidolia.
8. Boer Deng, "Plant Collections Left in the Cold by Cuts," *Nature* 523, no. 16 (July 2015), https://doi.org/10.1038/523016a.
9. J. J. Tewksbury, J. G. T. Anderson, J. D. Bakker, T. J. Billo, P. W. Dunwiddie, M. J. Groom, S. E. Hampton, S. G. Herman, D. J. Levey, N. J. Machnicki, C. Martínez del Rio, M. E. Power, K. Rowell, A. K. Salomon, L. Stacey, S. C. Trombulak, and T. A. Wheeler, "Natural History's Place in Science and Society," *Bioscience* 64, no. 4 (2014): 300–310.
10. Jennifer Frazer, "Natural History Is Dying, and We Are All the Losers," *The Artful Amoeba* (blog), *Scientific American*, June 20, 2014, https://blogs.scientificamerican.com/artful-amoeba/natural-history-is-dying-and-we-are-all-the-losers/.

6. COOL TOYS

1. I. A. Rahman and S. Y. Smith, "Virtual Paleontology: Computer-Aided Analysis of Fossil Form and Function," *Journal of Paleontology* 88, no. 4 (2014): 633–35.
2. P. L. Falkingham, "Acquisition of High Resolution 3D Models Using Free, Open-Source, Photogrammetric Software," *Palaeontologia Electronica* 15, no. 1 (2012): article 15.1.1T, https://doi.org/10.26879/264.

3. T. Boodhoo, T. Neier, and J. Matthews, "digitalquarryproject," October 11, 2015, http://carnegiequarry.com/digital-quarry/.
4. J. B. Pruitt, N. G. Clement, and L. Tapanila, "Laser and Structured Light Scanning to Acquire 3-D Morphology," *Paleontological Society Papers* 22 (2017): 57–69.
5. C. A. Grant, B. J. MacFadden, P. Antonenko, and V. J. Perez, "3-D Fossils for K–12 Education: A Case Example Using the Giant Extinct Shark *Carcharocles megalodon*," *Paleontological Society Papers* 22 (2017): 197–209.
6. C. Clabby, "Paleontology's X-Ray Excavations," *American Scientist* 102, no. 5 (2014): 386, https://www.americanscientist.org/article/paleontologys-x-ray-excavations.You can download 3-D images at the Digital Atlas of Ancient Life at https://www.digital atlasofancientlife.org/vc/.
7. Derek Briggs and his colleagues have used this method to study three-dimensionally preserved fossils from the Silurian of Herefordshire, United Kingdom. For example, see D. E. G. Briggs, D. J. Siveter, D. J. Siveter, M. D. Sutton, R. J. Garwood, and D. Legg, "Silurian Horseshoe Crab Illuminates the Evolution of Arthropod Limbs," *Proceedings of the National Academy of Sciences* 109, no. 39 (2012): 15702–5.
8. R. Racicot, "Fossil Secrets Revealed: X-Ray CT Scanning and Applications in Paleontology," *Paleontological Society Papers* 22 (2017): 21–38.
9. U. Bergmann, P. L. Manning, and R. A. Wogelius, "Chemical Mapping of Paleontological and Archeological Artifacts with Synchrotron X-Rays," *Annual Review of Analytical Chemistry* 5 (July 2012): 361–89.
10. C. Soriano, M. Archer, D. Azar, P. Creaser, X. Delclòs, H. Godthelp, S. Hand, A. Jones, A. Nel, D. Néraudeau, J. Ortega-Blanco, R. Pérez-de la Fuente, V. Perrichot, E. Saupe, M. S. Kraemer, and P. Tafforeau, "Synchrotron X-Ray Imaging of Inclusions in Amber," *Comptes Rendus Palevol* 9, no. 6 (2010): 361–68.
11. G. D. Cody, N. S. Gupta, D. E. G. Briggs, A. L. D. Kilcoyne, R. E. Summons, F. Kenig, R. E. Plotnick, and A. C. Scott, "Molecular Signature of Chitin-Protein Complex in Paleozoic Arthropods," *Geology* 39, no. 3 (2011): 255–58.
12. A. O. Marshall and C. P. Marshall, "Vibrational Spectroscopy of Fossils," *Palaeontology* 58, no. 2 (2015): 201–11.
13. J. Wiemann, T.-R. Yang, and M. A. Norell, "Dinosaur Egg Colour Had a Single Evolutionary Origin," *Nature* 563, no. 7732 (2018): 555.
14. W. T. Fraser, A. C. Scott, A. E. S. Forbes, I. J. Glasspool, R. E. Plotnick, F. Kenig, and B. H. Lomax, "Evolutionary Stasis of Sporopollenin Biochemistry Revealed by Unaltered Pennsylvanian Spores," *New Phytologist* 196, no. 2 (2012): 397–401.
15. K. R. Moore, T. Bosak, F. A. Macdonald, D. J. G. Lahr, S. Newman, C. Settens, and S. B. Pruss, "Biologically Agglutinated Eukaryotic Microfossil from Cryogenian Cap Carbonates," *Geobiology* 15, no. 4 (2017): 499–515.
16. D. E. G., Briggs and R. E. Summons, "Ancient Biomolecules: Their Origins, Fossilization, and Role in Revealing the History of Life," *Bioessays* 36, no. 5 (2014): 482–90.
17. T. Clements, A. Dolocan, P. Martin, M. A. Purnell, J. Vinther, and S. E. Gabbott, "The Eyes of *Tullimonstrum* Reveal a Vertebrate Affinity," *Nature* 532, no. 7600 (2016): 500–503; V. E. McCoy, E. E. Saupe, J. C. Lamsdell, L. G. Tarhan, S. McMahon, S. Lidgard, P. Mayer, C. D. Whalen, C. Soriano, L. Finney, S. Vogt, E. G. Clark, R. P. Anderson, H.

Petermann, E. R. Locatelli, and D. E. G. Briggs, "The 'Tully Monster' Is a Vertebrate," *Nature* 532, no. 7600 (2016): 496–99. A paper that disagrees with the vertebrate interpretation is L. Sallan, S. Giles, R. S. Sansom, J. T. Clarke, Z. Johanson, I. J. Sansom, and P. Janvier, "The 'Tully Monster' Is Not a Vertebrate: Characters, Convergence and Taphonomy in Palaeozoic Problematic Animals," *Palaeontology* 60, no. 2 (2017): 149–57. And one that confirms the initial results is V. E. McCoy, J. Wiemann, J. C. Lamsdell, C. D. Whalen, S. Lidgard, P. Mayer, H. Petermann, and D. E. G. Briggs, "Chemical Signatures of Soft Tissues Distinguish Between Vertebrates and Invertebrates from the Carboniferous Mazon Creek Lagerstätte of Illinois," *Geobiology* 18, no. 5 (September 2020): 560–65.

18.   The ratio is also dependent on the original proportions in the ocean water. As continental glaciers grow, the oceans become enriched in the heavy 18O isotope. Relatively high values of 18O indicate both cold temperatures and the presence of ice sheets.

19.   As discussed in the *Omnivore's Dilemma*, the carbon isotopes of the teeth of most American's show a heavy influence of C4 plants because we eat so much corn sugar and other corn derived products.

20.   We do not have direct measurements of climate or ocean chemistry from before the development of scientific instruments. Instead, we need to find things that respond in a predictable way to variations in such properties as temperature, rainfall, salinity, or pH. These are called *proxies*. Examples include oxygen isotopes, tree rings, or the abundance of certain types of microfossils or tree pollen. A great deal of work has gone into calibrating these proxies with the variables of interest.

21.   M. Henehan, G. L. Foster, H. C. Bostock, R. Greenop, B. J. Marshall, and P. A. Wilson, "A New Boron Isotope-pH Calibration for *Orbulina universa*, with Implications for Understanding and Accounting for 'Vital Effects,'" *Earth and Planetary Science Letters* 454 (2016): 282–92.

22.   To download Past, see Natural History Museum, UiO, "Past 4 – The Past of the Future," https://www.nhm.uio.no › research › infrastructure › past.

7. BIG DATA AND THE BIG PICTURE

1.   A wonderful history of the paleobiological revolution has been written by David Sepkoski, Jack Sepkoski's son. His descriptions of many of the key figures are spot on. Although I knew most of them, I was more an observer than a participant in the events he describes. D. Sepkoski, *Rereading the Fossil Record: The Growth of Paleobiology as an Evolutionary Discipline* (Chicago: University of Chicago Press, 2012).

2.   S. J. Gould, "The Promise of Paleobiology as a Nomothetic, Evolutionary Discipline," *Paleobiology* 6 (1980): 96–118.

3.   In fact, in his 1978 presidential address to the Paleontological Society, Raup had stated: "I feel in a somewhat strange position today as the first president of The Society who has never described a species." D. M. Raup, "Approaches to the Extinction Problem," *Journal of Paleontology* 52, no. 3 (May 1978): 517–23, at 517.

4.   There is a huge literature on all of these subjects. See A. I. Miller, "Conversations About Phanerozoic Global Diversity," *Paleobiology* 26 (2000): 53–73.

5. D. M. Raup and J. J. Sepkoski, "Periodicity of Extinctions in the Geologic Past," *Proceedings of the National Academy of Sciences* 81 (1984): 801–5.

6. J. J. Sepkoski, "A Factor Analytic Description of the Phanerozoic Marine Fossil Record," *Paleobiology* 7 (1981): 36–53.

7. J. J. Sepkoski, "10 Years in the Library—New Data Confirm Paleontological Patterns," *Paleobiology* 19 (1993): 43–51.

8. These include taxonomic information from the Sepkoski compendium, the Evolution of Terrestrial Ecosystems Database, which was developed by Kay Behrensmeyer and her colleagues at the Smithsonian and included a structure for entering geological, ecological, and preservational data; a database on Ordovician fossils that Miller had been working on; databases on North America mammal taxonomy and distributions that Alroy had produced; and locality-based faunal lists compiled by Sepkoski.

9. A full description of the Analytical Paleobiology Workshop, its history, and course materials can be found at http://www.analytical.palaeobiology.de/; a description and list of participants of the earlier iterations is at http://fossilworks.org/?page=workshop.

10. The instructors are not paid; they only receive travel expenses. Phil Novack-Gottshall says that they "do it purely to help train the next generation of students."

11. R. Plotnick and P. Wagner, "Round Up the Usual Suspects: Common Genera in the Fossil Record and the Nature of Wastebasket Taxa," *Paleobiology* 32, no. 1 (2006): 126–46.

12. J. Alroy, M. Aberhan, D. J. Bottjer, M. Foote, F. T. Fürsich, P. J. Harries, et al., "Phanerozoic Trends in the Global Diversity of Marine Invertebrates," *Science* 321 (2008): 97–100.

13. This episode of Earth history was popularized by Steve Gould's bestseller: S. J. Gould, *Wonderful Life* (New York: Norton, 1989).

14. D. H. Erwin, M. Laflamme, S. M. Tweedt, E. A. Sperling, D. Pisani, and K. J. Peterson, "The Cambrian Conundrum: Early Divergence and Later Ecological Success in the Early History of Animals," *Science* 334 (2011): 1091–97.

15. E. A. Sperling, C. J. Wolock, A. S. Morgan, B. C. Gill, M. Kunzmann, G. P. Halverson, et al., "Statistical Analysis of Iron Geochemical Data Suggests Limited Late Proterozoic Oxygenation," *Nature* 523 (2015): 451–54.

16. D. M. Raup and J. J. Sepkoski, "Mass Extinctions in the Marine Fossil Record," *Science* 215, no. 4539 (1982): 1501–3.

17. R. K. Bambach, A. H. Knoll, and S. C. Wang "Origination, Extinction, and Mass Depletions of Marine Diversity," *Paleobiology* 30, no. 4 (2004): 522–42: A. L. Stigall, "Invasive Species and Biodiversity Crises: Testing the Link in the Late Devonian," *PLoS One* 5, no. 12 (2010): e15584.

18. C. R. Marshall, "Five Palaeobiological Laws Needed to Understand the Evolution of the Living Biota," *Nature Ecology & Evolution* 1, no. 6 (2017): 165.

8. THE ENDS OF THE WORLDS AS WE KNOW THEM

1. E. O. Wilson, "The Biological Diversity Crisis," *Bioscience* 35, no. 11 (1985): 700–706, quote at 703.

2. In their 1995 book, Richard Leakey and Roger Lewin suggested that E. O. Wilson coined the phrase "Sixth Extinction" in 1986. R. Leakey and R. Lewin, *The Sixth Extinction: Patterns of Life and the Future of Humankind* (New York: Doubleday, 1995).

3. The number of mass extinctions may not be five. The number of episodes of elevated extinction or drops in marine biodiversity may be as many as eighteen, but only three of these could count as episodes of true mass extinction. R. K. Bambach, "Phanerozoic Biodiversity Mass Extinctions," *Annual Review of Earth and Planetary Sciences* 34 (2006): 127–55.

4. L.W. Alvarez, W. Alvarez, F. Asaro, and H. V. Michel, "Extraterrestrial Cause for the Cretaceous-Tertiary Extinction—Experimental Results and Theoretical Interpretation," *Science* 208 (1980): 1095–1108; A. R. Hildebrand, G. T. Penfield, D. A. Kring, M. Pilkington, A. Camargo, S. B. Jacobsen, et al., "Chicxulub Crater—A Possible Cretaceous Tertiary Boundary Impact Crater on the Yucatan Peninsula, Mexico," *Geology* 19 (1991): 867–71.

5. B. Schoene, M. P. Eddy, K. M. Samperton, C. B. Keller, G. Keller, T. Adatte, and S. F. R. Khadri, "U-Pb Constraints on Pulsed Eruption of the Deccan Traps Across the End-Cretaceous Mass Extinction," *Science* 363, no. 6429 (2019): 862–66; C. J. Sprain, P. R. Renne, L. Vanderkluysen, K. Pande, S. Self, and T. Mittal, "The Eruptive Tempo of Deccan Volcanism in Relation to the Cretaceous-Paleogene Boundary," *Science* 363, no. 6429 (2019): 866–70.

6. M. J. Henehan, A. Ridgwell, E. Thomas, S. Zhang, L. Alegret, D. N. Schmidt, J. W. B. Rae, J. D. Witts, N. H. Landman, S. E. Greene, B. T. Huber, J. R. Super, N. J. Planavsky, and P. M. Hull, "Rapid Ocean Acidification and Protracted Earth System Recovery Followed the End-Cretaceous Chicxulub Impact," *Proceedings of the National Academy of Sciences* 116, no. 45 (2019): 22500–22504; A. A. Chiarenza, A. Farnsworth, P. D. Mannion, D. J. Lunt, P. J. Valdes, J. V. Morgan, and P. A. Allison, "Asteroid Impact, Not Volcanism, Caused the End-Cretaceous Dinosaur Extinction," *Proceedings of the National Academy of Sciences* 117, no. 29 (2020): 17084–93.

7. S. P. S. Gulick, T. J. Bralower, J. Ormö, B. Hall, K. Grice, B. Schaefer, S. Lyons, K. H. Freeman, J. V. Morgan, N. Artemieva, P. Kaskes, S. J. de Graaff, M. T. Whalen, G. S. Collins, S. M. Tikoo, C. Verhagen, G. L. Christeson, P. Claeys, M. J. L. Coolen, S. Goderis, K. Goto, R. A. F. Grieve, N. McCall, G. R. Osinski, A. S. P. Rae, U. Riller, J. Smit, V. Vajda, and A. Wittmann, "The First Day of the Cenozoic," *Proceedings of the National Academy of Sciences* 116, no. 39 (2019): 19342–51.

8. R. A. DePalma, J. Smit, D. A. Burnham, K. Kuiper, P. L. Manning, A. Oleinik, P. Larson, F. J. Maurrasse, J. Vellekoop, M. A. Richards, L. Gurche, and W. Alvarez, "A Seismically Induced Onshore Surge Deposit at the KPg Boundary, North Dakota," *Proceedings of the National Academy of Sciences* 116, no. 17 (2019): 8190–99.

9. Judith Viorst, *Alexander and the Terrible, Horrible, No Good, Very Bad Day* (New York: Simon & Schuster, 1987).

10. Because birds are phylogenetically a group of dinosaurs, this phrase was developed to make the point that dinosaurs are, therefore, technically not extinct. But I think the "nonavian" part could be considered understood when we talk about the extinct dinosaurs; after all, insects are phylogenetically a group of crustaceans, but we don't

feel compelled to say "nonhexapod" crustaceans when discussing crabs, lobsters, shrimp, etc.

11. A. Weil and J. W. Kirchner, "Diversity on the Rebound," *Nature Ecology & Evolution* 3, no. 6 (2019): 873–74.

12. C. M. Lowery, T. J. Bralower, J. D. Owens, F. J. Rodríguez-Tovar, H. Jones, J. Smit, M. T. Whalen, P. Claeys, K. Farley, S. P. S. Gulick, J. V. Morgan, S. Green, E. Chenot, G. L. Christeson, C. S. Cockell, M. J. L. Coolen, L. Ferrière, C. Gebhardt, K. Goto, D. A. Kring, J. Lofi, R. Ocampo-Torres, L. Perez-Cruz, A. E. Pickersgill, M. H. Poelchau, A. S. P. Rae, C. Rasmussen, M. Rebolledo-Vieyra, U. Riller, H. Sato, S. M. Tikoo, N. Tomioka, J. Urrutia-Fucugauchi, J. Vellekoop, A. Wittmann, L. Xiao, K. E. Yamaguchi, and W. Zylberman, "Rapid Recovery of Life at Ground Zero of the End-Cretaceous Mass Extinction," *Nature* 558 (2018): 288–91.

13. S. A. Alvarez, S. J. Gibbs, P. R. Bown, H. Kim, R. M. Sheward, and A. Ridgwell, "Diversity Decoupled from Ecosystem Function and Resilience During Mass Extinction Recovery," *Nature* 574, no. 7777 (2019): 242–45.

14. T. R. Lyson, I. M. Miller, A. D. Bercovici, K. Weissenburger, A. J. Fuentes, W. C. Clyde, J. W. Hagadorn, M. J. Butrim, K. R. Johnson, R. F. Fleming, R. S. Barclay, S. A. Maccracken, B. Lloyd, G. P. Wilson, D. W. Krause, and S. G. B. Chester, "Exceptional Continental Record of Biotic Recovery After the Cretaceous–Paleogene Mass Extinction," *Science* 366, no. 6468 (2019): 977–83.

15. D. J. Field, A. Bercovici, J. S. Berv, R. Dunn, D. E. Fastovsky, T. R. Lyson, V. Vajda, and J. A. Gauthier, "Early Evolution of Modern Birds Structured by Global Forest Collapse at the End-Cretaceous Mass Extinction," *Current Biology* 28, no. 11 (2018): 1825–32; D. M. Grossnickle, S. M. Smith, and G. P. Wilson, "Untangling the Multiple Ecological Radiations of Early Mammals," *Trends in Ecology & Evolution* 34, no. 10 (2019): 936–49.

16. M. J. Benton, *When Life Nearly Died: The Greatest Mass Extinction of All Time* (New York: Thames & Hudson, 2003); D. H. Erwin, *Extinction: How Life on Earth Nearly Ended 250 Million Years Ago* (Princeton, NJ: Princeton University Press, 2006).

17. In 1979, Dave Raup estimated the size of the marine extinction as between 88 percent and 96 percent; the upper figure is the one most often quoted. Steven Stanley reduced that estimate to 81 percent of marine species. D. M. Raup, "Size of the Permo-Triassic Bottleneck and Its Evolutionary Implications," *Science* 206, no. 4415 (1979): 217–18; S. M. Stanley, "Estimates of the Magnitudes of Major Marine Mass Extinctions in Earth History," *Proceedings of the National Academy of Sciences* 113, no. 42 (2016): E6325–34.

18. S. D. Burgess, J. D. Muirhead, and S. A. Bowring, "Initial Pulse of Siberian Traps Sills as the Trigger of the End-Permian Mass Extinction," *Nature Communications* 8, no. 1 (2017): 164.

19. S. D. Burgess, S. Bowring, and S.-z. Shen, "High-Precision Timeline for Earth's Most Severe Extinction," *Proceedings of the National Academy of Sciences* 111, no. 9 (2014): 3316–21.

20. H. Nowak, E. Schneebeli-Hermann, and E. Kustatscher, "No Mass Extinction for Land Plants at the Permian–Triassic Transition," *Nature Communications* 10, no. 1 (2019):

384; R. A. Gastaldo, "Plants Escaped an Ancient Mass Extinction," *Nature* 567, no. 7746 2019): 38–39.

21. C. R. Fielding, T. D. Frank, S. McLoughlin, V. Vajda, C. Mays, A. P. Tevyaw, A. Winguth, C. Winguth, R. S. Nicoll, M. Bocking, and J. L. Crowley, "Age and Pattern of the Southern High-Latitude Continental End-Permian Extinction Constrained by Multiproxy Analysis," *Nature Communications* 10, no. 1 (2019): 385; C. Mays, V. Vajda, T. D. Frank, C. R. Fielding, R. S. Nicoll, A. P. Tevyaw, and S. McLoughlin, "Refined Permian–Triassic Floristic Timeline Reveals Early Collapse and Delayed Recovery of South Polar Terrestrial Ecosystems," *GSA Bulletin* 132, no. 7–8 (2019): 1489–1513.

22. R. A. Gastaldo, S. L. Kamo, J. Neveling, J. W. Geissman, C. V. Looy, and A. M. Martini, "The Base of the *Lystrosaurus* Assemblage Zone, Karoo Basin, Predates the End-Permian Marine Extinction," *Nature Communications* 11, no. 1 (2020): 8.

23. S. E. Grasby, B. Beauchamp, and J. Knies, "Early Triassic Productivity Crises Delayed Recovery from World's Worst Mass Extinction," *Geology* 44, no. 9 (2016): 779–82.

24. R. Dirzo, H. S. Young, M. Galetti, G. Ceballos, N. J. B. Isaac, and B. Collen, "Defaunation in the Anthropocene," *Science* 345, no. 6195 (2014): 401–6.

25. Y. M. Bar-On, R. Phillips, and R. Milo, "The Biomass Distribution on Earth," *Proceedings of the National Academy of Sciences* 115, no. 25 (2018): 6506–11.

26. C. T. Darimont, C. H. Fox, H. M. Bryan, and T. E. Reimchen, "The Unique Ecology of Human Predators," *Science* 349, no. 6250 (2015): 858–60.

27. D. J. McCauley, M. L. Pinsky, S. R. Palumbi, J. A. Estes, F. H. Joyce, and R. R. Warner, "Marine Defaunation: Animal Loss in the Global Ocean," *Science* 347, no. 6219 (2015): 1255641.

28. A. D. Barnosky, N. Matzke, S. Tomiya, G. O. U. Wogan, B. Swartz, T. B. Quental, C. Marshall, J. L. McGuire, E. L. Lindsey, K. C. Maguire, B. Mersey, and E. A. Ferrer, "Has the Earth's Sixth Mass Extinction Already Arrived?," *Nature* 471, no. 7336 (2011): 51–57.

29. R. E. Plotnick, F. A. Smith, and S. K. Lyons, "The Fossil Record of the Sixth Extinction," *Ecology Letters* 19, no. 5 (2016): 546–53.

30. The "gold standard" for the status of species is the International Union for the Conservation of Nature (IUCN) Red List of Threatened Species, https://www.iucnredlist.org/. Coverage for terrestrial vertebrates is excellent, less so for invertebrates and marine organisms.

31. Many of the points raised here were discussed by Doug Erwin, a specialist on the Permian extinction, in a 2017 interview. See P. Brannen, "Earth Is Not in the Midst of a Sixth Mass Extinction," *The Atlantic*, June 13, 2017, https://www.theatlantic.com /science/archive/2017/06/the-ends-of-the-world/529545/.

32. A. D. Barnosky, P. L. Koch, R. S. Feranec, S. L. Wing, and A. B. Shabel, "Assessing the Causes of Late Pleistocene Extinctions on the Continents," *Science* 306, no. 5693 (2004): 70–75.

33. Y. Malhi, C. E. Doughty, M. Galetti, F. A. Smith, J.-C. Svenning, and J. W. Terborgh, "Megafauna and Ecosystem Function from the Pleistocene to the Anthropocene," *Proceedings of the National Academy of Sciences* 113, no. 4 (2016): 838–46.

34. D. J. Meltzer, "Overkill, Glacial History, and the Extinction of North America's Ice Age Megafauna," *Proceedings of the National Academy of Sciences* 117, no. 46 (2020):

28555; J. Alroy, "A Multispecies Overkill Simulation of the End-Pleistocene Megafaunal Mass Extinction," *Science* 292, no. 5523 (2001): 1893–96; A. D. Barnosky, P. L. Koch, R. S. Feranec, S. L. Wing, and A. B. Shabel, "Assessing the Causes of Late Pleistocene Extinctions on the Continents," *Science* 306, no. 5693 (2004): 70–75.

35. S. K. Lyons, K. L. Amatangelo, A. K. Behrensmeyer, A. Bercovici, J. L. Blois, M. Davis, W. A. DiMichele, A. Du, J. T. Eronen, J. Tyler Faith, G. R. Graves, N. Jud, C. Labandeira, C. V. Looy, B. McGill, J. H. Miller, D. Patterson, S. Pineda-Munoz, R. Potts, B. Riddle, R. Terry, A. Tóth, W. Ulrich, A. Villaseñor, S. Wing, H. Anderson, J. Anderson, D. Waller, and N. J. Gotelli, "Holocene Shifts in the Assembly of Plant and Animal Communities Implicate Human Impacts," *Nature* 529, no. 7584 (2016): 80–83; F. A. Smith, S. M. Elliott, and S. K. Lyons, "Methane Emissions from Extinct Megafauna," *Nature Geoscience* 3, no. 6 (2010): 374–75.

36. E. Ellis, M. Maslin, N. Boivin, and A. Bauer, "Involve Social Scientists in Defining the Anthropocene," *Nature* 540 (2016): 192–93.

37. L. Stephens, D. Fuller, N. Boivin, T. Rick, N. Gauthier, A. Kay, B. Marwick, C. G. Armstrong, C. M. Barton, T. Denham, K. Douglass, J. Driver, L. Janz, P. Roberts, J. D. Rogers, H. Thakar, M. Altaweel, A. L. Johnson, M. M. Sampietro Vattuone, M. Aldenderfer, S. Archila, G. Artioli, M. T. Bale, T. Beach, F. Borrell, T. Braje, P. I. Buckland, N. G. Jiménez Cano, J. M. Capriles, A. Diez Castillo, Ç. Çilingiroğlu, M. Negus Cleary, J. Conolly, P. R. Coutros, R. A. Covey, M. Cremaschi, A. Crowther, L. Der, S. di Lernia, J. F. Doershuk, W. E. Doolittle, K. J. Edwards, J. M. Erlandson, D. Evans, A. Fairbairn, P. Faulkner, G. Feinman, R. Fernandes, S. M. Fitzpatrick, R. Fyfe, E. Garcea, S. Goldstein, R. C. Goodman, J. Dalpoim Guedes, J. Herrmann, P. Hiscock, P. Hommel, K. A. Horsburgh, C. Hritz, J. W. Ives, A. Junno, J. G. Kahn, B. Kaufman, C. Kearns, T. R. Kidder, F. Lanoë, D. Lawrence, G.-A. Lee, M. J. Levin, H. B. Lindskoug, J. A. López-Sáez, S. Macrae, R. Marchant, J. M. Marston, S. McClure, M. D. McCoy, A. V. Miller, M. Morrison, G. Motuzaite Matuzeviciute, J. Müller, A. Nayak, S. Noerwidi, T. M. Peres, C. E. Peterson, L. Proctor, A. R. Randall, S. Renette, G. Robbins Schug, K. Ryzewski, R. Saini, V. Scheinsohn, P. Schmidt, P. Sebillaud, O. Seitsonen, I. A. Simpson, A. Sołtysiak, R. J. Speakman, R. N. Spengler, M. L. Steffen, M. J. Storozum, K. M. Strickland, J. Thompson, T. L. Thurston, S. Ulm, M. C. Ustunkaya, M. H. Welker, C. West, P. R. Williams, D. K. Wright, N. Wright, M. Zahir, A. Zerboni, E. Beaudoin, S. Munevar Garcia, J. Powell, A. Thornton, J. O. Kaplan, M.-J. Gaillard, K. Klein Goldewijk, and E. Ellis, "Archaeological Assessment Reveals Earth's Early Transformation Through Land Use," *Science* 365, no. 6456 (2019): 897–902.

38. For the indigenous footprint in the Americas and the impact of the European discovery of the New World, required reading are these two books: C. C. Mann, *1491: New Revelations of the Americas Before Columbus* (New York: Random House, 2005); and C. C. Mann, *1493: Uncovering the New World Columbus Created* (New York: Random House, 2011).

39. R. E. Plotnick and K. A. Koy, "The Anthropocene Fossil Record of Terrestrial Mammals," *Anthropocene* 29 (March 2020): 100233.

40. There are both strong proponents and opponents for the formal naming of the Anthropocene as part of the geological time scale. A full discussion of this issue would

be a book in itself! See J. Zalasiewicz, M. Williams, A. Haywood, and M. Ellis, "The Anthropocene: A New Epoch of Geological Time?," *Philosophical Transactions of the Royal Society A: Mathematical, Physical and Engineering Sciences* 369, no. 1938 (2011): 835–41; S. C. Finney and L. E. Edwards, "The 'Anthropocene' Epoch: Scientific Decision or Political Statement?," *GSA Today* 26, no. 3–4 (2016): 4–9.

41. R. A. Spicer, A. B. Herman, and E. M. Kennedy, "The Sensitivity of CLAMP to Taphonomic Loss of Foliar Physiognomic Characters," *PALAIOS* 20, no. 5 (2005): 429–38.

42. B. H. Lomax and W. T. Fraser, "Palaeoproxies: Botanical Monitors and Recorders of Atmospheric Change," *Palaeontology* 58, no. 5 (2015): 759–68; I. P. Montanez, J. C. McElwain, C. J. Poulsen, J. D. White, W. A. DiMichele, J. P. Wilson, G. Griggs, and M. T. Hren, "Climate, p_{CO_2} and Terrestrial Carbon Cycle Linkages During Late Palaeozoic Glacial-Interglacial Cycles," *Nature Geoscience* 9, no. 11 (2016): 824–28; J. C. McElwain, "Paleobotany and Global Change: Important Lessons for Species to Biomes from Vegetation Responses to Past Global Change," *Annual Review of Plant Biology* 69, no. 1 (2018): 761–87.

43. J. D. Hays, J. Imbrie, and N. J. Shackleton, "Variations in the Earth's Orbit: Pacemaker of the Ice Ages," *Science* 194, no. 4270 (1976): 1121–32.

44. T. Westerhold, N. Marwan, A. J. Drury, D. Liebrand, C. Agnini, E. Anagnostou, J. S. K. Barnet, S. M. Bohaty, D. De Vleeschouwer, F. Florindo, T. Frederichs, D. A. Hodell, A. E. Holbourn, D. Kroon, V. Lauretano, K. Littler, L. J. Lourens, M. Lyle, H. Pälike, U. Röhl, J. Tian, R. H. Wilkens, P. A. Wilson, and J. C. Zachos, "An Astronomically Dated Record of Earth's Climate and Its Predictability over the Last 66 Million Years," *Science* 369, no. 6509 (2020): 1383–87.

45. G. P. Dietl and K. W. Flessa, "Conservation Paleobiology: Putting the Dead to Work," *Trends in Ecology & Evolution* 26, no. 1 (2011): 30–37, quote at 30.

46. D. A. Fordham, S. T. Jackson, S. C. Brown, B. Huntley, B. W. Brook, D. Dahl-Jensen, M. T. P. Gilbert, B. L. Otto-Bliesner, A. Svensson, S. Theodoridis, J. M. Wilmshurst, J. C. Buettel, M. Canteri, M. McDowell, L. Orlando, J. Pilowsky, C. Rahbek, and D. Nogues-Bravo, "Using Paleo-Archives to Safeguard Biodiversity Under Climate Change," *Science* 369, no. 6507 (2020): eabc5654; W. Kiessling, N. B. Raja, V. J. Roden, S. T. Turvey, and E. E. Saupe, "Addressing Priority Questions of Conservation Science with Palaeontological Data," *Philosophical Transactions of the Royal Society B: Biological Sciences* 374, no. 1788 (2019): 20190222; A. D. Barnosky, E. A. Hadly, P. Gonzalez, J. Head, P. D. Polly, A. M. Lawing, J. T. Eronen, D. D. Ackerly, K. Alex, E. Biber, J. Blois, J. Brashares, G. Ceballos, E. Davis, G. P. Dietl, R. Dirzo, H. Doremus, M. Fortelius, H. W. Greene, J. Hellmann, T. Hickler, S. T. Jackson, M. Kemp, P. L. Koch, C. Kremen, E. L. Lindsey, C. Looy, C. R. Marshall, C. Mendenhall, A. Mulch, A. M. Mychajliw, C. Nowak, U. Ramakrishnan, J. Schnitzler, K. Das Shrestha, K. Solari, L. Stegner, M. A. Stegner, N. C. Stenseth, M. H. Wake, and Z. Zhang, "Merging Paleobiology with Conservation Biology to Guide the Future of Terrestrial Ecosystems," *Science* 355, no. 6325 (2017): eaah4787.

47. R. Lockwood and R. Mann, "A Conservation Palaeobiological Perspective on Chesapeake Bay Oysters," *Philosophical Transactions of the Royal Society B: Biological Sciences* 374, no. 1788 (2019): 20190209.

48. E. Kolbert, *The Sixth Extinction: An Unnatural History* (New York: Picador, 2014).

9. LESSONS FOR AND FROM THE LIVING

1. This approach was suggested by Robert Aller, a sedimentary geochemist then at the University of Chicago.

2. R. E. Plotnick, "Taphonomy of a Modern Shrimp: Implications for the Arthropod Fossil Record," *PALAIOS* 1 (1986): 286–93, doi:10.2307/3514691.

3. For a recent review, see M. A. Purnell, P. J. C. Donoghue, S. E. Gabbott, M. E. McNamara, D. J. E Murdock, and R. S. Sansom, "Experimental Analysis of Soft-Tissue Fossilization: Opening the Black Box," *Paleontology* 61 (2018): 317–23, doi:10.1111/pala.12360.

4. D. Erwin, *Paleontology After Gould*, May 5, 2017, http://www.extinctblog.org/extinct /2017/5/30/paleontology-after-gould. This essay is a necessary corrective to the idea that Steve Gould was the only important paleontological thinker of his time.

5. As Steve Holland put it in his 2016 presidential address to the Paleontological Society, "As paleontologists, we have an extraordinary data set at our disposal, and we have the expertise to understand it. We have something that no other field of biology has— time, deep time." S. M. Holland, "Presidential Address: Structure, Not Bias," *Journal of Paleontology* 91 (2017): 1315–17, https://doi.org/10.1017/jpa.2017.114. The robust structure of the fossil record allows testing of sophisticated hypotheses about the history of life, and our understanding of taphonomy promotes rather than inhibits the framing of these hypotheses.

6. J. J. Smith, E. Turner, A. Möller, R. M. Joeckel, and R. E. Otto, "First U-Pb Zircon Ages for Late Miocene Ashfall Konservat-Lagerstätte and Grove Lake Ashes from Eastern Great Plains, USA," *PLoS One* 13, no. 11 (2018): e0207103.

7. Lance Grande of the Field Museum has spent decades collecting and describing fossils from the Green River formation. His book on the area also has many stunning photographs. See L. Grande, *The Lost World of Fossil Lake: Snapshots from Deep Time* (Chicago: University of Chicago Press, 2013).

8. G. P. Dietl and P. H. Kelley, "The Fossil Record of Predator-Prey Arms Races: Coevolution and Escalation Hypotheses," *The Fossil Record of Predation: Paleontological Society Special Papers* 8 (2002): 353–74.

9. Conrad Labandeira (National Museum of Natural History) and his colleagues have published an identification guide to insect damage. See C. C. Labandeira, P. Wilf, K. R. Johnson, and F. Marsh, *Guide to Insect (and Other) Damage Types on Compressed Plant Fossils* (version 3.0) (Washington, DC: Smithsonian Institution, 2007).

10. E. D. Currano, "Ancient Bug Bites on Ancient Plants Record Forest Ecosystem Response to Environmental Perturbations," *Paleontological Society Papers* 19 (2013): 157–74.

11. J. H. Miller, A. K. Behrensmeyer, A. Du, S. K. Lyons, D. Patterson, A. Toth, A. Villasenor, E. Kanga, and D. Reed, "Ecological Fidelity of Functional Traits Based on Species Presence-Absence in a Modern Mammalian Bone Assemblage (Amboseli, Kenya)," *Paleobiology* 40, no. 4 (2014): 560–83; J. T. Faith and A. K. Behrensmeyer, "Changing Patterns of Carnivore Modification in a Landscape Bone Assemblage, Amboseli Park, Kenya," *Journal of Archaeological Science* 33, no. 12 (2006): 1718–33.

12. E. N. Powell, G. M. Staff, W. R. Callender, K. A. Ashton-Alcox, C. E. Brett, K. M. Parsons-Hubbard, S. E. Walker, and A. Raymond, "Taphonomic Degradation of

Molluscan Remains During Thirteen Years on the Continental Shelf and Slope of the Northwestern Gulf of Mexico," *Palaeogeography Palaeoclimatology Palaeoecology* 312, no. 3–4 (2011): 209–32.

13. Susan Kidwell has spent much of her career studying the thick deposits known as shell beds. See S. M. Kidwell, "Time-Averaging and Fidelity of Modern Death Assemblages: Building a Taphonomic Foundation for Conservation Palaeobiology," *Palaeontology* 56, no. 3 (2013): 487–522.

14. X. Ma, G. D. Edgecombe, X. Hou, T. Goral, and N. J. Strausfeld, "Preservational Pathways of Corresponding Brains of a Cambrian Euarthropod," *Current Biology* 25, no. 22 (2015): 2969–75.

15. I like the definition of a fossil used by Behrensmeyer et al.: "any non-living, biologically generated trace or material that paleontologists study as part of the record of life." A. K. Behrensmeyer, S. M. Kidwell, and R. A. Gastaldo, "Taphonomy and Paleobiology," *Paleobiology* 26, no. 4 (2000): 103–47, quote at 104.

16. D. E. G. Briggs and R. E. Summons, "Ancient Biomolecules: Their Origins, Fossilization, and Role in Revealing the History of Life," *BioEssays* 36, no. 5 (2014): 482–90.

17. J. Vinther, "A Guide to the Field of Palaeo Colour," *BioEssays* 37, no. 6 (2015): 643–56.

18. J. Wiemann, T.-R. Yang, P. N. Sander, M. Schneider, M. Engeser, S. Kath-Schorr, C. E. Müller, and P. M. Sander, "Dinosaur Origin of Egg Color: Oviraptors Laid Blue-Green Eggs," *PeerJ* 5 (2017): e3706.

19. The muscles must act against the ligament to keep the shell closed. When a clam dies, the muscles relax and the shell opens. Never buy a clam with gaping shells at the fish market!

20. L. Selby, *When Your Anatomy Professor Is a Paleontologist*, The Do, February 13, 2016, http://thedo.osteopathic.org/2016/02/when-your-anatomy-professor-is-a-paleontologist/.

21. R. E. Plotnick and T. K. Baumiller, "Invention by Evolution: Functional Analysis in Paleobiology," *Paleobiology* 26, no. 4 (2000): 305–23.

22. S. M. Stanley, *Relation of Shell Form to Life Habits in the Bivalvia (Mollusca)* (Lawrence, KS: Geological Society of America, 1970).

23. For example, oysters live in massive clumps on the surface and are said to be epifaunal, whereas clams live buried in the sediment and are infaunal. The thin shells and single massive muscle of scallops is associated with them being able to detach from the surface and swim to avoid predators (the muscle is what you eat).

24. Vogel was an amazing professor and the best textbook writer I know. For a taste of his writing, see S. Vogel, *Cats' Paws and Catapults* (New York: Norton, 1998). LaBarbera is also a fantastic teacher and was a member of my doctoral committee.

25. R. E. Plotnick and J. Bauer, "Crinoids Aweigh: Experimental Biomechanics of *Ancyrocrinus* Holdfasts," in *Experimental Approaches to Understanding Fossil Organisms*, ed. D. I. Hembree, B. F. Platt, and J. J. Smith (Dordrecht: Springer, 2014), 1–18.

26. R. D. C. Bicknell, J. A. Ledogar, S. Wroe, B. C. Gutzler, W. H. Watson, and J. R. Paterson, "Computational Biomechanical Analyses Demonstrate Similar Shell-Crushing Abilities in Modern and Ancient Arthropods," *Proceedings of the Royal Society B: Biological Sciences* 285, no. 1889 (2018): 20181935; J. A. Nyakatura, K. Melo,

T. Horvat, K. Karakasiliotis, V. R. Allen, A. Andikfar, E. Andrada, P. Arnold, J. Lauströer, J. R. Hutchinson, M. S. Fischer, and A. J. Ijspeert, "Reverse-Engineering the Locomotion of a Stem Amniote," *Nature* 565, no. 7739 (2019): 351–55; I. A. Rahman, "Computational Fluid Dynamics as a Tool for Testing Functional and Ecological Hypotheses in Fossil Taxa," *Palaeontology* 60, no. 4 (2017): 451–59.

27. A. J. Martin, *Life Traces of the Georgia Coast* (Bloomington: Indiana University Press, 2013). Tony has also written books on dinosaur trace fossils and the evolution of burrowing.

28. An entire book can be written about the introduction of cladistics as a methodology, the often hostile reception it received, and the internal strife within the proponents of the method. For a recent discussion of the "cladistic wars," see B. Sterner and S. Lidgard, "Moving Past the Systematics Wars," *Journal of the History of Biology* 51, no. 4 (2017): 1–37.

29. C. R. Marshall, "Five Palaeobiological Laws Needed to Understand the Evolution of the Living Biota," *Nature Ecology & Evolution* 1 (2017): 0165.

30. M. Pagel, "Evolutionary Trees Can't Reveal Speciation and Extinction Rates," *Nature* 580, no. 7804 (2020): 461–62, quote at 462.

31. J. A. Cunningham, A. G. Liu, S. Bengtson, and P. C. J. Donoghue, "The Origin of Animals: Can Molecular Clocks and the Fossil Record Be Reconciled?," *BioEssays* 39, no. 1 (2017): 1–12; E. A. Sperling and R. G. Stockey, "The Temporal and Environmental Context of Early Animal Evolution: Considering All the Ingredients of an 'Explosion,'" *Integrative and Comparative Biology* 58, no. 4 (2018): 605–22.

32. G. M. Erickson, "Assessing Dinosaur Growth Patterns: A Microscopic Revolution," *Trends in Ecology & Evolution* 20, no. 12 (2005): 677–84.

33. T. Dobzhansky, "Nothing in Biology Makes Sense Except in Light of Evolution," *American Biology Teacher* 35, no. 3 (March 1973): 125–29.

10. THE EDUCATION OF A PALEONTOLOGIST

1. Nicholette Zeliadt, "Profile of David Jablonski," *Proceedings of the National Academy of Sciences* 110, no. 26 (2013): 10467–69.

2. "UCMP Salutes Bill Clemens," *UCMP News*, January 2003, 4, https://ucmp.berkeley .edu/museum/ucmp_news/2003/1-03/clemens4.html.

3. For a long time, there was a degree-granting Department of Paleontology at the University of California, Berkeley, which in 1989 merged with other biology departments to form what is now the Department of Integrative Biology. The University of Alberta in Edmonton offers an undergraduate specialization in paleontology.

4. Many geology departments are rethinking this requirement because not everyone is physically capable of the necessary grueling fieldwork. The International Association for Geoscience Diversity has been at the forefront of addressing this issue.

5. S. E. Widnall, "AAAS Presidential Lecture: Voices from the Pipeline," *Science* 241, no. 4874 (1988): 1740–45.

11. LIVING IN THE REAL WORLD

1. NEA-AFT, "The Truth About Tenure in Higher Education" (online brochure), n.d., https://diversityinhighereducation.com/articles/The-Truth-About-Tenure-in-Higher-Education.

2. Unfortunately, this has become more common. A talented colleague of mine, with tenure, lost his position at an Illinois university that was in financial distress. He was fortunate to find a position elsewhere.

3. K. W. Flessa and D. M. Smith, "Paleontology in Academia: Recent Trends and Future Opportunities," *Paleontology in the 21st Century* (Frankfurt: Paleo21, 1997).

4. R. E. Plotnick, "A Somewhat Fuzzy Snapshot of Employment in Paleontology in the United States," *Palaeontologia Electronica* 11, no. 1 (2008): E1–3.

12. THE FACE OF PALEONTOLOGY

1. R. E. Plotnick, A. L. Stigall, and I. Stefanescu, "Evolution of Paleontology: Long-Term Gender Trends in an Earth Science Discipline," *GSA Today* 24 (2014): 44–45.

2. The idea that she inspired the tongue-twister "She sells sea shells on the sea shore" has been debunked. S. Winick, "She Sells Seashells and Mary Anning: Metafolklore with a Twist," *Folklore Today* (blog), Library of Congress, July 26, 2017, https://blogs.loc.gov/folklife/2017/07/she-sells-seashells-and-mary-anning-metafolklore-with-a-twist/.

3. The Paleontological Research Institute, whose long-time director Katherine Palmer was one of these proud few, has a wonderful website, Daring to Dig: Women in American Paleontology (http://www.daringtodig.com/), and has developed a traveling exhibition that rescues these women from obscurity.

4. In "Alumnae, Graduate School, Yale University, 1894–1920," Yale University.

5. P. A. Cohen, A. Stigall, and C. Topaz, "A Gender Analysis of the Paleontological Society: Trends, Gaps, and a Way Forward," poster presented at the 11th North American Paleontological Convention, June 23–27, 2019, Riverside, California.

6. C. B. de Wet, G. M. Ashley, and D. P. Kegel, "Biological Clocks and Tenure Timetables: Restructuring the Academic Timeline," *GSA Today* 12, no. 11 (2002): 1–7.

7. W. M. Williams and S. J. Ceci, "When Scientists Choose Motherhood," *American Scientist* 100 (2012): 138–45.

8. N. H. Wolfinger, "For Female Scientists, There's No Good Time to Have Children," *The Atlantic*, July 29, 2013, https://www.theatlantic.com/sexes/archive/2013/07/for-female-scientists-theres-no-good-time-to-have-children/278165/.

9. M. Urry, "Women in (European) Astronomy," in *Formation and Evolution of Cosmic Structures: Reviews in Modern Astronomy*, vol. 21, ed. S. Röser (Weinheim, Germany: Wiley-VCH, 2010), 249–61.

10. A. L. Stigall, "The Paleontological Society 2013: A Snapshot in Time," *Priscum* 20 (2013): 1–4.

11. Since 2000 a third of the winners received their PhD at the University of Chicago. Here is the breakdown: Chicago (7); Harvard (3); Kansas (2). All of the following have

one winner: Columbia University, University of Arizona, Cornell, Penn State, Leicester (UK), Edinburgh (UK), University of California Berkeley, University of California Riverside, University of Southern California.

12. I brought a graduate student from a Mexican American family to a GSA meeting. At one point he asked, "Where are the people who look like me?"

13. C. E. Wilson, "Race and Ethnicity of U.S. Citizen Geoscience Graduate Students and Postdoctoral Appointees, 2016," American Geosciences Institute, updated October 31, 2018, https://www.americangeosciences.org/geoscience-currents/race-and-ethnicity-us -citizen-geoscience-graduate-students-and-postdoctoral-appointees.

14. C. Wilson, "Status of Recent Geoscience Graduates 2017," 1-49: American Geosciences Institute, 2018, https://www.americangeosciences.org/sites/default/files/ExitSurvey_2017 _Online_041018.pdf.

15. S. O'Connell and M. A. Holmes, "Obstacles to the Recruitment of Minorities in the Geosciences: A Call to Action," GSA Today 21, no. 6 (2011): 52–54.

16. P. J. Stokes, R. Levine, and K. W. Flessa, "Why Are There So Few Hispanic Students in Geoscience?," GSA Today 24, no. 1 (2014): 52–53.

17. J. E. Huntoon, C. Tanenbaum, and J. Hodges, "Increasing Diversity in the Geosciences," Eos, March 9, 2015.

18. The use of the acronym BIPOC (Black, Indigenous, people of color) has gained popularity but is not universally welcomed.

19. Paleontological Society, "The Paleontological Society Statement on Anti-Racism," 2020, https://www.paleosoc.org/assets/docs/Paleontological-Society-How-to-Be-an-Anti -Racism-Ally.pdf.

20. "A Call to Action for an Anti-Racist Science Community from Geoscientists of Color: Listen, Act, Lead," No Time for Silence, https://notimeforsilence.org/.

21. Riley Black wrote a compelling commentary on this. See R. Black, "It's Time for the Heroic Male Paleontologist Trope to Go Extinct," Slate, April 3, 2019, https:// slate.com/technology/2019/04/what-the-new-yorker-dinosaur-story-gets-wrong .html.

13. THE THIRD REVIEWER

1. S. J. Gould, "Allometric Fallacies and the Evolution of Gryphaea: A New Interpretation Based on White's Criterion of Geometric Similarity," Evolutionary Biology 6 (1972): 91–118.

2. D. M. Raup, "A Kill Curve for Phanerozoic Marine Species," Paleobiology 17, no. 1 (Winter 1991): 37–48.

3. In November 2020, the publisher Springer Nature announced an open access fee of 9,500 euros (about $11,000) for Nature and thirty-two other journals under the Nature umbrella. Holly Else, "Nature Journals Reveal Terms of Landmark Open-Access Option," Nature, November 24, 2020, https://www.nature.com/articles/d41586-020 -03324-y.

14. CONFERRING, CONVERSING, AND OTHERWISE HOBNOBBING

1. L. Guertin, "Want to Be More Inclusive? Stop Making Geology Conferences About the Beer," *GeoEd Trek* (blog), December 8, 2019, https://blogs.agu.org/geoedtrek/2019 /12/08/inclusive-agu-conferences/.

2. Geocognition Research Laboratory, "Limiting Alcohol During Professional Events," July 2, 2020, https://geocognitionresearchlaboratory.com/2020/07/02/limiting-alcohol -during-professional-events/; Geological Society of America, "Geological Society of America Announces 2020–2021 Officers and Councilors," https://www.geosociety.org /GSA/News/pr/2020/20-20.aspx.

3. A marvelous organization that addresses issues of accessibility in the geosciences is the International Association for Geoscience Diversity at https://theiagd.org/.

15. FIGHTING OVER SCRAPS

1. M. Mitchell, M. Leachman, and K. Masterson, "Funding Down, Tuition Up: State Cuts to Higher Education Threaten Quality and Affordability at Public Colleges," Center on Budget and Policy Priorities, updated August 15, 2016, http://hdl.handle .net/10919/97764.

2. S. Vogel, "Macroscope: Academically Correct Biological Science," *American Scientist* 86, no. 6 (1998): 504–6.

3. Dena Smith, personal communication, September 2017

4. When I started out, the requirement was fifteen physical copies received by the deadline. This created more than a little tension and some conflicts around the photocopy machine.

5. Patricia Kelley was an NSF program director, a "rotator," from 1990 to 1992. Writing in the now defunct magazine *American Paleontologist*, she reported that she was able to fund 34 percent of proposals, but only by "cutting award budgets to the minimum necessary to complete the research." The success rate for 2001 comes from a 2008 presentation by the NSF program directors at a townhall meeting.

6. You can see the official NSF guide to proposal preparation at https://www.nsf.gov /pubs/policydocs/pappg20_1/index.jsp.

7. University of Florida paleontologist Bruce MacFadden has published a guide for the perplexed, see B. MacFadden, *Broader Impacts of Science on Society* (Cambridge: Cambridge University Press, 2019), doi:10.1017/9781108377577.

8. For eighteen years, until his death in 2015, the permanent SGP program director was Richard Lane; it is currently Dena Smith.

9. For example, 108 paleontologists from thirty countries met at the Senckenberg Museum, Frankfurt, Germany, in September 1997, to participate in a conference titled "Paleontology in the 21st Century." As was the case for many such meetings, there were a lot of bold plans (six goals for 2020) but no follow-through. Similarly, Karl Flessa convened an NSF-sponsored PS workshop on "Geobiology and the Earth Sciences in the Next Decade," at the Smithsonian in March 1999 that identified four research themes. Patricia Kelley, personal communication, August 29, 2019.

10. S. M. Stanley, et al., eds., *Geobiology of Critical Intervals* (Lawrence, KS: Paleontological Society, 1997). It should be noted that although the definition of geobiology used in this document explicitly includes paleontology, more recent usage of geobiology has tended to restrict it to the interactions of microbial life with the environment, including over geological time scales.

11. "DETELON Workshop Report," 2010, organized by Dave Bottjer and Doug Erwin.

12. E. Pennisi, "NSF's Huge Ecological Observatory Is Open for Business: But Tensions Remain," *Science*, August 29, 2019, https://www.sciencemag.org/news/2019/08/nsf-s -huge-ecological-observatory-open-business-tensions-remain.

13. *EAR to the Ground* (newsletter), NSF EAR division, fall 2012.

14. P. D. Polly, J. J. Head, and D. L. Fox, eds., "Earth-Life Transitions: Paleobiology in the Context of Earth System Evolution," *Paleontological Society Papers*, vol. 21, October 2015.

15. These key people included Jack Hess (GSA), Howard Harper (SEPM), Martin Perlmutter (Chevron), and Philip Gingerich (Paleontological Society). I would like to thank Judy Parrish, Howard Harper, Dena Smith, Martin Perlmutter, and Lisa Park Boush for discussing STEPPE with me.

16. D. M. Smith and D. Iler, "STEPPE: Earth's Past, Our Future," *The Sedimentary Record* (March 2015): 4–9.

17. I am grateful to Patricia Kelley for her detailed notes on these workshops and initiatives. The image of paleontology as an old, outdated science persists; combating it is one of my incentives for this book.

18. Total funding for the PDBD is about $4 million, this compares to about $100 million for IDigBio.

19. C. R. Marshall, S. Finnegan, E. C. Clites, P. A. Holroyd, N. Bonuso, C. Cortez, E. Davis, G. P. Dietl, P. S. Druckenmiller, R. C. Eng, C. Garcia, K. Estes-Smargiassi, A. Hendy, K. A. Hollis, H. Little, E. A. Nesbitt, P. Roopnarine, L. Skibinski, J. Vendetti, and L. D. White, "Quantifying the Dark Data in Museum Fossil Collections as Palaeontology Undergoes a Second Digital Revolution," *Biology Letters* 14, no. 9 (2018).

16. THIS LAND IS YOUR LAND, YOUR FOSSIL IS MY FOSSIL

1. The skeleton of STAN was sold by the Black Hills Institute and was part of the separation of Neal Larson from the institute. An October 20, 2020, press release stated in part: "The auction was the end result of a lawsuit filed by Neal Larson in 2015 to liquidate the Institute's assets. In 2018, a judge ordered the Institute to auction the bones of STAN™, and pay all of the proceeds to Neal Larson for his interest in the company. . . . We were saddened to learn that the winner of the auction was probably not a museum, but we are hopeful that the new owner will eventually put STAN™ on display so the public will be able to continue to see and study this awesome original skeleton."

2. J. Pickrell, "Carnivorous-Dinosaur Auction Reflects Rise in Private Fossil Sales," *Nature*, June 6, 2018, https://www.nature.com/articles/d41586-018-05299-3.

3. For some reason, the Field Museum prefers the name in all caps: "SUE."

4. For the most detailed account, see S. Fiffer, *Tyrannosaurus SUE: The Extraordinary Saga of Largest, Most Fought Over T. Rex Ever Found* (New York: W. H. Freeman, 2001). For Peter Larson's own account, see P. Larson and K. Donnan, *Rex Appeal: The Amazing Story of Sue, the Dinosaur That Changed Science, the Law, and My Life* (Invisible Cities Press, 2004). For the Field Museum's perspective, see L. Grande, *Curators: Behind the Scenes of Natural History Museums* (Chicago: University of Chicago Press, 2017). The *Nova* episode in 1997, "The Curse of *T.rex*," provides a balanced discussion of the issues. A transcript can be read at http://www.pbs.org/wgbh/nova/transcripts/2408trex.html/. For a Native American perspective, see L. Bradley, *Dinosaurs and Indians: Paleontology Resource Dispossession from Sioux Lands* (Denver: Outskirts Press, 2014). For a detailed review of the legal aspects, see A. M. Dussias, "Science, Sovereignty, and the Sacred Text: Paleontological Resources and Native American Rights," *Maryland Law Review* 55, no. 5 (1996): 84–159. Todd Douglas Miller directed the slanted 2013 film *Dinosaur 13*. For a highly critical review of the film, see D. Lessem, "Don't Believe the Anti-Government Tale Spun by This New Dinosaur Documentary," *Slate*, August 22, 2014, https://slate.com/culture/2014/08/dinosaur-13-review-movie-about-peter-larson-spins-a-bogus-tale.html. In *Curators*, Grande says that "[the 2013 film *Dinosaur 13*] is an exploitative attempt to show unbroken admiration of the BHI as the David vs. Goliath underdog at the expense of everyone else" (114). The Society of Vertebrate Paleontology also published a critical response.

5. For a thorough account of the case, see P. Williams, "Bones of Contention," *New Yorker*, January 20, 2013, https://www.newyorker.com/magazine/2013/01/28/bones-of-contention-paige-williams. Paige has now expanded this into a book: P. Williams, *The Dinosaur Artist: Obsession, Betrayal, and the Quest for Earth's Ultimate Trophy* (New York: Hachette, 2018).

6. L. Pyne, "The Second Life of Mongolian Fossils," *The Atlantic*, December 27, 2017, https://www.theatlantic.com/science/archive/2017/12/second-life-of-mongolian-fossils/548558/.

7. For a history of the specimen, see J. L. Franzen, P. D. Gingerich, J. Habersetzer, J. H. Hurum, W. von Koenigswald, and B. H. Smith, "Complete Primate Skeleton from the Middle Eocene of Messel in Germany: Morphology and Paleobiology," *PLoS One* 4, no. 5 (2009): e5723.

8. C. Zimmer, "Science Held Hostage," *Discover Magazine*, May 21, 2019, https://www.discovermagazine.com/planet-earth/science-held-hostage.

9. E. L. Simons, F. Ankel-Simons, P. S. Chatrath, R. S. Kay, B. Williams, J. G. Fleagle, D. L. Gebo, C. K. Beard, M. Dawson, I. Tattersall, and K. D. Rose, "Outrage at High Price Paid for a Fossil," *Nature* 460, no. 7254 (2009): 456.

10. B. J. Switek, "Ancestor or Adapiform? *Darwinius* and the Search for Our Early Primate Ancestors," *Evolution: Education and Outreach* 3, no. 3 (2010): 468–76.

11. There has been a great deal of news coverage on the dueling dinosaurs. For a story sympathetic to Phipps, see M. Sager, "Will the Public Ever Get to See the 'Dueling Dinosaurs'?," *Smithsonian Magazine*, July/August 2017, https://www.smithsonianmag.com/science-nature/public-ever-see-dueling-dinosaurs-180963676/. For two stories that cover the protracted legal battle, see W. Cornwall, "Court Rules

'Dueling Dinos' Belong to Landowners, in a Win for Science," *Science Magazine*, May 22, 2020, https://www.sciencemag.org/news/2020/05/court-rules-dueling-dinos-belong-landowners-win-science; and J. Feinstein and A. Wernick, "'Dueling Dinos' Set Off a Long Legal Battle and a Scientific Debate," *The World*, April 9, 2020, https://www.pri.org/stories/2020-04-09/dueling-dinos-set-long-legal-battle-and-scientific-debate. Phipps is featured in the Discovery Channel series *Dino Hunters*, billed as the "Montana rancher Clayton Phipps and Wyoming fossil hunter Mike Harris buck the academic status quo by putting their money on the line, traveling the earth, and turning out groundbreaking finds that've set the scientific world on fire." The Society for Vertebrate Paleontology sharply criticized the show for its "tone-deaf promotion of the idea that paleontology is the sole prerogative of white men" and "that this series highlights and glamourizes the sale of vertebrate fossils to the highest bidder." See Society of Vertebrate Paleontology (letter), June 11, 2020, http://vertpaleo.org/GlobalPDFS/SVP_Discovery_Dino_Hunter_2020-revised.aspx.

12. M. Greshko, "'Dueling Dinosaurs' Fossil, Hidden from Science for 14 Years, Could Finally Reveal Its Secrets," *National Geographic*, November 17, 2020, https://www.nationalgeographic.com/science/2020/11/dueling-dinosaurs-fossil-finally-set-to-reveal-secrets/; J. Shaffer, "'Dueling Dinosaurs': World's 1st Complete Tyrannosaurus rex, Triceratops, Raleigh Bound," *News and Observer*, November 17, 2020, https://www.newsobserver.com/news/local/article246605893.html.

13. Grande, *Curators*; for a balanced discussion of private ownership of fossils, see H. Meijer, "To Collect or Not to Collect: Are Fossil-Hunting Laws Hurting Science," *The Guardian*, July 27, 2016, https://www.theguardian.com/science/2016/jul/27/to-collect-or-not-to-collect-are-fossil-hunting-laws-hurting-science.

14. D. M. Martill, H. Tischlinger, and N. R. Longrich, "A Four-Legged Snake from the Early Cretaceous of Gondwana," *Science* 349, no. 6246 (2015): 416–19.

15. C. Gramling, "Update: Controversial 'Four-Legged Snake' May Be Ancient Lizard Instead," *Science Magazine*, November 11, 2016, https://www.sciencemag.org/news/2016/11/update-controversial-four-legged-snake-may-be-ancient-lizard-instead. As of 2016, the fossil is again on museum display.

16. Society of Vertebrate Paleonteology, letter to the editors, April 21, 2020, https://vertpaleo.org/wp-content/uploads/2021/01/SVP-Letter-to-Editors-FINAL.pdf.

17. Society of Vertebrate Paleontology, "Bylaws," article 12, section 6, http://vertpaleo.org/wp-content/uploads/2021/01SVP-Bylaws-a. The Paleontological Society, "Code of Fossil Collecting" is less prescriptive, see https://www.paleosoc.org/code-of-fossil-collecting.

18. K. Shimada, P. J. Currie, E. Scott, and S. S. Sumida, "The Greatest Challenge to 21st Century Paleontology: When Commercialization of Fossils Threatens the Science," *Palaeontologia Electronica* 17, no. 1 (2014): 1E. 4.

19. Founded in 1997, *Palaeontologia Electronica* (https://palaeo-electronica.org/) was the first peer-reviewed paleontology journal to be solely online, free, and open access.

20. R. E. Plotnick, "Out of the Mainstream: Fossil Collecting in the 21st Century," *Palaeontologica Electronica* 14, no. 1 (2011), https://palaeo-electronica.org/2011_1/commentary/mainstream.htm.

21. Neal Larson has since left BHI; the sale of "STAN" was part of the separation agreement.

22. P. L. Larson and D. Russell, "The Benefits of Commercial Fossil Sales to 21st Century Paleontology," *Palaeontologia Electronica* 17, no. 1 (2014): 2E, 1–7.

23. N. L. Larson, W. Stein, M. Triebold, and G. Winters, "What Commercial Fossil Dealers Contribute to the Science of Paleontology," *Journal of Paleontological Sciences* 7 (Fall 2014). Full disclosure: a version of the commentary had also been originally submitted to PE. As editor, I rejected it due to its tone and lack of advancing the discussion. This version is much better.

24. This is an oft-made point, including in the Paleontological Society "Code of Fossil Collecting," https://www.paleosoc.org/code-of-fossil-collecting.

25. Association of Applied Paleontogical Sciences (AAPS), "AAPS Commercial Paleontology Code of Ethics," https://www.aaps.net/ethics.htm.

26. J. Picktrell, "How Fake Fossils Pervert Paleontology [Excerpt]," *Scientific American*, November 15, 2014, https://www.scientificamerican.com/article/how-fake-fossils-pervert-paleontology-excerpt/.

27. Grande, *Curators*, 44.

28. J. A. Catalani, "Contributions by Amateur Paleontologists in 21st Century Paleontology," *Palaeontologia Electronica* 17, no. 2 (2014): 3E. A response from the paleoart community also appeared, see M. P. Witton, D. Naish, and J. Conway, "Commentary: State of the Palaeoart," *Palaeontologia Electronica* 17, no. 3 (2014): 5E.

29. Lauer Foundation for Paleontology, Science, and Education, lauerfoundationpse.org.

30. V. L. Santucci and J. M. Ghist, "Fossil Cycad National Monument: A History from Discovery to Deauthorization," *Dakoterra* 6 (2014): 82–93, quote at 92.

31. Language of the act is available at Paleontological Resources Preservation Act.pdf (blm.gov). Consult the Society of Vertebrate Paleontology website for a discussion and FAQ from the perspective of the SVP at http://vertpaleo.org.

32. Julia Brunner of the National Park Service kindly explained the convoluted history of implementing the PRPA (07/02/2020).

33. For final regulations by the Forest Service, including comments and responses, see Forest Service, "Paleontological Resources Preservation," *Federal Register*, U.S. National Archives, April 17, 2015, https://www.federalregister.gov/documents/2015/04/17/2015-08483/paleontological-resources-preservation.

34. For the proposed rules for the Bureau of Interior agencies, see Bureau of Land Management, Bureau of Reclamation, U.S. National Park Service, U.S. Fish and Wildlife Service, and the U.S. Department of the Interior, "Paleontological Resources Preservation," *Federal Register*, U.S. National Archives, December 7, 2016, https://www.federalregister.gov/documents/2016/12/07/2016-29244/paleontological-resources-preservation. According to David Polly (November 2020), the Office of Management and Budget rejected the proposed regulations. A different set of rules govern the collection of petrified wood on BLM lands; see Bureau of Land Management, U.S. Department of the Interior, "Recreation Activities," https://www.blm.gov/basic/rockhounding.

35.   B. F. Dattilo, C. E. Brett, D. L. Meyer, R. L. Freeman, B. Hunda, S. M. Holland, A. L. Stigall, B. Deline, C. D. Sumrall, and M. A. Wilson, "Non-academic Paleontologists Are Essential to the Survival of Paleontology: Lessons from the Cincinnati School Geological Society of America," *Abstracts with Programs* 47 (2015): 582.

36.   For an interview with Linda, see "Amateur Spotlight: Linda McCall," interview by Jennifer Bauer, *myFOSSIL*, https://www.myfossil.org/amateur-spotlight-linda-mccall/. For additional information about Linda, see Malia Wollan, "How to Find Fossils," *New York Times*, September 10, 2019, https://www.nytimes.com/2019/09/10/magazine/how-to-find-fossils.html.

37.   Linda McCall: Comment on proposed Rule on Paleontological Resources Preservation Act of 2009 (PRPA), January 8, 2017, (personal communication, June 22, 2021).

38.   Of course, there is the expropriation of the lands themselves. It is becoming increasingly common to explicitly recognize whose land we currently occupy; this book is being written on what used to be Potawatomi territory.

39.   PBS, "Examining the History of Fossil Collection from Native American Lands," *Prehistoric Road Trip* (series), WTTW.com, https://interactive.wttw.com/prehistoric-road-trip/stops/examining-the-history-of-fossil-collection-from-native-american-lands.

40.   L. Bradley, *Dinosaurs and Indians: Paleontology Resource Dispossession from Sioux Lands* (Denver, CO: Outskirts Press, 2014). A paper by Adrienne Mayor covers some of the same territory but adds additional examples, both in the United States and abroad, see A. Mayor, "Fossils in Native American Lands Whose Bones, Whose Story? Fossil Appropriations Past and Present," presentation at the History of Science Society, Washington, D.C., 2007.

41.   A. M. Dussias, "Science, Sovereignty, and the Sacred Text: Paleontological Resources and Native American Rights," *Maryland Law Review* 55, no. 5 (1996): 84–159, quote at 156.

42.   Standing Rock Sioux Tribal Council, "Title 38, Paleontology Resource Code," approved February 3, 2015, https://www.standingrock.org/Titles/.

43.   "Proclamation 6920—Establishment of the Grand Staircase–Escalante National Monument," September 18, 1996, https://www.govinfo.gov/content/pkg/WCPD-1996-09-23/pdf/WCPD-1996-09-23-Pg1788.pdf.

44.   David Polly, as cited in A. Reese, "The Bones of Bears Ears," *Science* 363 (2019): 218–20.

45.   Office of the Press Secretary, The White House, "Presidential Proclamation—Establishment of the Bears Ears National Monument," December 28, 2016, https://obama whitehouse.archives.gov/the-press-office/2016/12/28/proclamation-establishment-bears-ears-national-monument. For images of some of the archeological sites, see J. Fox, L. Tierney, S. Blanchard, and G. Florit, "What Remains of Bears Ears," *Washington Post*, design by J. Crump, April 2, 2019, https://www.washingtonpost.com/graphics/2019/national/bears-ears/.

46.   Associated Press, "Native American Tribes to Fight President's Order on National Monuments," *Washington Post*, December 5, 2017, https://www.washingtonpost.com/lifestyle/kidspost/native-american-tribes-to-fight-presidents-order-on-national

-monuments/2017/12/05/5666be7c-d9e5-11e7-b1a8-62589434a581_story.html; Patagonia, "Hey, How's That Lawsuit Against the President Going?," https://www.patagonia .com/stories/hey-hows-that-lawsuit-against-the-president-going/story-72248.html.

47. A. Reese, "Scientists Sue to Protect Utah Monument—and Fossils That Could Rewrite Earth's History," *Science Magazine*, January 17, 2019, https://www.sciencemag.org /news/2019/01/scientists-sue-protect-utah-monument-and-fossils-could-rewrite -earth-s-history; National Parks Conservation Association, "Plaintiff Organizations in Bears Ears and Grand Staircase—Escalante National Monument Cases Denounce Administration's Final Management Plans" (press release), February 6, 2020.

48. myFossil, "Fossil Parks," https://www.myfossil.org/fossil-parks/. There is also an international PaleoParks program that focuses on protecting key fossil sites; see J. Lipps, "PaleoParks: Our Paleontological Heritage Protected and Conserved in the Field Worldwide," chapter 1 in *PaleoParks—The Protection and Conservation of Fossil Sites Worldwide*, ed. J. H. Lipps and B. R. C. Granier (Brest, France: Carnets de Géologie / Notebooks on Geology, 2009).

49. M. C. Morton, "The Snowmastodon Project: Mammoths and Mastodons Lived the High Life in Colorado," *Earth Magazine*, December 9, 2015, https://www.earthmagazine.org /article/snowmastodon-project-mammoths-and-mastodons-lived-high-life-colorado.

50. P. C. Murphey, G. E. Knauss, L. H. Fisk, T. A. Deméré, R. E. Reynolds, K. C. Trujillo, and J. J. Strauss, "A Foundation for Best Practices in Mitigation Paleontology," *Dakoterra* 6 (2014): 243–85; P. C. Murphey, G. E. Knauss , L. H. Fisk, T. A. Deméré, and R. E. Reynolds, "Best Practices in Mitigation Paleontology," *Proceedings of the San Diego Society of Natural History* 47 (2019): 1–43.

51. K. R. Johnson and R. Troll, *Cruisin' the Fossil Coastline* (Golden, CO: Fulcrum, 2018), 22.

17. FOR THE LOVE OF FOSSILS

1. R. G. Johnson and E. S. Richardson Jr., "Pennsylvanian Invertebrates of the Mazon Creek Area, Illinois: The Morphology and Affinities of *Tullimonstrum*," *Fieldiana: Geology* 12, no. 8 (1969): 119–49.

2. A biography of Langford written by his son is available online: George Langford Jr., "The Life of George Langford Sr., 1876–1964," https://www.georgesbasement.com /McKennaProcessCo/HTMLfiles/GeorgeLangfordSrbiographyByGLJr.htm.

3. J. A. Catalani, "Contributions by Amateur Paleontologists in 21st Century Paleontology," *Palaeontologia Electronica* 17, no. 2 (2014): 3E.

4. ESCONI publications include G. Langford, *The Wilmington Coal Flora from a Pennsylvanian Deposit in Will County, Illinois* (Downers Grove, IL: Esconi Associates, 1958); G. Langford, *The Wilmington Coal Fauna: And Additions to the Wilmington Coal Flora from a Pennsylvanian Deposit in Will County, Illinois* (Downers Grove, IL: Esconi Associates, 1963); J. Wittry, *The Mazon Creek Fossil Flora* (Downers Grove, IL: Earth Science Club of Northern Illinois, 2006); J. Wittry, *The Mazon Creek Fossil Fauna* (Downers Grove, IL: Earth Science Club of Northern Illinois, 2012).

5. For a huge posthumous monograph on fossil scorpions, see E. N. Kjellesvig-Waering, "A Restudy of the Fossil Scorpionida of the World," *Palaeontographica Americana* 55 (1986): 1–285.

6. In 2016, Samuel Ciurca was honored for his contributions by receiving the Strimple Award of the Paleontological Society.

7. Jennifer Bauer, "Amateur Spotlight: Linda McCall"(interview), *myFOSSIL*, https://www.myfossil.org/amateur-spotlight-linda-mccall/. In 2020, McCall was awarded the Strimple Award of the Paleontological Society. She was the first solo woman so honored.

8. B. Lefebvre, R. Lerosey-Aubril, T. Servais, and P. Van Roy, introduction to "The Fezouata Biota: An Exceptional Window on the Cambro-Ordovician Faunal Transition," *Palaeogeography, Palaeoclimatology, Palaeoecology* 460 (2016): 1–6.

9. B. J. MacFadden, L. Lundgren, K. Crippen, B. A. Dunckel, and S. Ellis, "Amateur Paleontological Societies and Fossil Clubs, Interactions with Professional Paleontologists, and Social Paleontology in the United States," *Palaeontologia Electronica* 19, no. 2 (2016): 1E.

10. The upper part of the Ordovician in North America is known as the Cincinnatian.

11. A detailed description of the Cincinnati School can be found in D. L. Meyer and R. A. Davis, *A Sea Without Fish: Life in the Ordovician Sea of the Cincinnati Region* (Bloomington: Indiana University Press, 2009).

12. Roy Plotnick, "The Old Boy's Network: Academic Genealogy and Gender," *Medium*, July 22, 2018, https://medium.com/@plotnick/the-old-boys-network-academic-genealogy-and-gender-eeod5881abb5.

13. Fostering Opportunities for Synergistic STEM with Informal Learners. For those not familiar with education and grant-speak: "STEM" is Science, Technology, Education, and Math and "Informal Learners" are those who learn outside of the classroom. Sometimes you write the acronym first and then figure out what words fit it. This is also what I call a meta-acronym because it is an acronym that contains an acronym.

14. Lauer Foundation, https://www.lauerfoundationpse.org/education

15. Microvertebrates are tiny vertebrate fossils, such as the bones and teeth of rodent-sized animals. They often have to be laboriously sieved from sediments and examined under a microscope. J. E. Peterson, R. P. Scherer, and K. M. Huffman, "Methods of Microvertebrate Sampling and Their Influences on Taphonomic Interpretations," *PALAIOS* 26, no. 1/2 (2011): 81–88.

16. A full list of winners can be found at https://paleo.memberclicks.net/past-awardees.

17. FossilGuy.com, www.fossilguy.com

18. Riley Black, personal communication, February 17, 2020.

18. PRESENTING THE PAST

1. After one production, the stage manager of the play took me aside and told me that I should wear a lab coat so I would "look like a scientist." My crabby response was a simple no. "I am the only scientist you know, and I don't wear a lab coat!"

2. M. Varelas, R. Plotnick, D. Wink, Q. Fan, and Y. Harris, "Inquiry and Connections in Integrated Science Content Courses for Elementary Education Majors," *Journal of College Science Teaching* 37, no. 5 (2008): 40–47; R. Plotnick, M. Varelas, and Q. Fan, "Physical World: An Integrated Earth Science, Astronomy and Physics Course for Elementary Education Majors," *Journal of Geoscience Education* 57, no. 2 (2009): 152–58.

3. For a detailed description of education and outreach, see B. J. MacFadden, *Broader Impacts of Science on Society* (Cambridge: Cambridge University Press, 2019).

4. R. C. Moore, C. G. Lalicker, and A. G. Fischer, *Invertebrate Fossils* (New York: McGraw-Hill, 1952). It is old, but still useful.

5. I learned paleontology from the revised 1968 edition: J. R. Beerbower, *Search for the Past: An Introduction to Paleontology* (London: Pearson, 1968).

6. D. M. Raup and S. M. Stanley, *Principles of Paleontology* (San Francisco: W. H. Freeman, 1971). A second edition was published in 1978, and a third and totally revised version, authored by former Raup students Arnie Miller and Mike Foote, came out in 2006. A. I. Miller and M. Foote, *Principles of Paleontology*, 3rd ed. (San Francisco: W. H. Freeman, 2006).

7. *Teaching Paleontology in the 21st Century* (workshop), Cornell University and the Paleontological Research Institute, Ithica, New York, July 30–August 4, 2009, https://serc .carleton.edu/NAGTWorkshops/paleo/workshop09/program.html. Published as M. M. Yacobucci and Rowan Lockwood, eds., *Teaching Paleontology in the 21st Century*, Paleontological Society Special Publication Vol. 12: Paleontological Society, 2012. This was one of a series of cutting-edge professional development programs for current and future geoscience faculty, organized by NAGT. I participated in a workshop on course development and ended up revamping how I taught.

    The 2018 short course contributions are published in the online series Elements of Paleontology. https://www.cambridge.org/core/what-we-publish/elements/elements-of -paleontology

8. R. Lockwood, P. A. Cohen, M. D. Uhen, and K. Ryker, *Utilizing the Paleobiology Database to Provide Educational Opportunities for Undergraduates* (Cambridge: Cambridge University Press, 2018).

9. In a traditional science course, lectures are done in the classroom and assignments at home. To a first approximation, in a flipped course the lecture is watched online at home and classroom time is used for assignments and exercises with the instructor present. For some courses, such as the statistics course I long taught, this sounds ideal.

10. Teaching is far more effective if we understand what conceptions and misconceptions students bring to the classroom. M. M. Yacobucci, *Confronting Prior Conceptions in Paleontology Courses* (Cambridge: Cambridge University Press, 2018).

11. Steve Semken defines *sense of place* as all the meanings and attachments held by an individual or a group for a place. The sense of place of students can be leveraged to make teaching science more effective. See S. Semken, "Sense of Place and Place-Based Introductory Geoscience Teaching for American Indian and Alaska Native Undergraduates," *Journal of Geoscience Education* 53 (2005): 149–57; C. C. Visaggi, *Equity, Culture, and Place in Teaching Paleontology: Student-Centered Pedagogy for Broadening Participation* (Cambridge: Cambridge University Press, 2020).

12. Achieve, Next Generation Science Standards, accessed July 15, 2021, https://www .nextgenscience.org/.

13. In 2019, Sarah Sheffield won the Biggs Award for Excellence in Earth Science Teaching from the Geoscience Education Division of the Geological Society of America.

14. Time Scavengers, https://timescavengers.blog/

15. I am thinking of you, Animal Planet.

16. To learn more about the three-part PBS series *Prehistoric Road Trip*, go to https:// www.pbs.org/show/prehistoric-road-trip/.

17. Burpee Museums, "Expeditions 2021," https://burpee.org/expeditions-2020. Unfortunately, the pandemic canceled the 2020 field season.

19. THOSE WHO DO NOT KNOW THE PAST

1. Bill Nye (The Science Guy) debated the young Earth creationist Ken Ham on the question "Is Creation a Viable Model of Origins?," February 2014, at the Creation Museum in Petersburg, Kentucky. I agree with those who felt this gave undeserved publicity and credibility to Ham and his ideas.

2. There are many excellent books laying out the case for evolution. Here are some authored by paleontologists: N. D. Newell, *Creation and Evolution: Myth or Reality?* (New York: Columbia University Press, 1982); N. Eldredge, *The Triumph of Evolution and the Failure of Creationism* (New York: Freeman, 2000): and D. R. Prothero, *Evolution: What the Fossils Say and Why It Matters* (New York: Columbia University Press, 2017), in which Don describes his experience "debating" creationist Duane Gish and other tactics used by opponents of evolution. Any of the books by Steve Gould laying out the case for evolution are good, for example, S. J. Gould, *Hen's Teeth and Horse's Toes: Further Reflections in Natural History* (New York: Norton, 1983).

   Anthropologist Eugenie Scott, of the National Center for Science Education, is a leading proponent for teaching evolution, see E. C. Scott, *Evolution vs. Creationism: An Introduction* (Westport, CT: Greenwood, 2009); and E. C. Scott and G. Branch, eds., *Not in Our Classrooms* (Boston: Beacon, 2006).

   For a recent analysis of the arguments used by creationists in the context of education, see R. M. Barnes, R. A. Church, and S. Draznin-Nagy, "The Nature of the Arguments for Creationism, Intelligent Design, and Evolution," *Science & Education* 26 (2017): 27–47. For an excellently written, albeit depressing review on climate change, see E. Kolbert, *Field Notes from a Catastrophe: Man, Nature, and Climate Change* (New York: Bloomsbury, 2006). And for a revealing look at how climate deniers operate, see N. Oreskes and E. Conway, *Merchants of Doubt* (London: Bloomsbury, 2010); there is also a 2014 documentary based on this book.

3. During the 2009 North American Paleontological Convention, a group of seventy attendees went to the recently opened Creation Museum (North American Paleontological Convention 2009, "Visit to the Creation Museum," http://napc2009.org /creationmuseum; Kenneth Chang, "Paleontology and Creationism Meet but Don't Mesh," *New York Times*, June 29, 2009, https://www.nytimes.com/2009/06/30/science/30muse .html). I refused to go, not wanting them to benefit from the price of my admission.

4. Museum of the Earth, *Changing Climate: Our Future, Our Choice*, https://www.museum oftheearth.org/changing-climate.
5. Paleontological Research Institution, *Digital Atlas of Ancient Life*, https://www.digital atlasofancientlife.org/.
6. Paleontological Research Institution, *The Teacher-Friendly Guide to Climate Change*, https://www.priweb.org/science-education-programs-and-resources/teacher-friendly -guide-to-climate-change.
7. University of California, Berkeley, *Understanding Evolution*, https://evolution.berkeley .edu/evolibrary/home.php.
8. https://ugc.berkeley.edu/
9. You can download a copy of the interactive graphic at https://undsci.berkeley.edu /lessons/pdfs/complex_flow_posterh.pdf.
10. An important paper that discusses the philosophical foundations of geology and helps dispel what has been called "physics envy" is R. Frodeman, "Geological Reasoning: Geology as an Interpretive and Historical Science," *Geological Society of America Bulletin* 107, no. 8 (1995): 960–968.
11. U.S. National Center for Science Education (NCSE), https://ncse.ngo/.
12. The vertebrate paleontologist Kevin Padian (UC Berkeley) was an expert witness at the trial and is a former president of NCSE's board of directors.
13. Paleontological Society, *The Paleontological Society Position Statement: Evolution*, https://www.paleosoc.org/index.php?option=com_content&view=article&id=73: evolution.
14. The SVP statement on evolution can be accessed through https://vertpaleo.org/wp -content/uploads/2021/01/On-Evolution.pdf.
15. S. J. Gould, *Rocks of Ages: Science and Religion in the Fullness of Life* (New York: Ballantine, 1999); S. J. Gould, "Nonoverlapping Magisteria," *Natural History* 106 (March 1997): 16–22, http://www.blc.arizona.edu/courses/schaffer/449/Gould%20Nonoverlapping%20Magisteria.htm.
16. P. H. Kelley, "Studying Evolution and Keeping the Faith," *Geotimes* 45, no. 12 (2000): 22–25, quote at 25.
17. The Clergy Letter Project, http://www.theclergyletterproject.org/rel_evol_sun.htm.
18. The Clergy Letter Project, *2020 Evolution Weekend*, http://theclergyletterproject.org /rel_evolution_weekend_2020.html.
19. Sinai and Synapses, https://sinaiandsynapses.org/. For a book that reviews many scientific issues, including evolution, from a Jewish perspective, see R. L. Price, *When Judaism Meets Science* (Eugene, OR: Wipf & Stock, 2019).
20. Although the most visible creationists in the United States are Christians, there are also creationists among Orthodox Jews and Muslims.

20. DRAWING ON THE PAST

1. Another influential muralist is Jay Matternes, whose work adorned the walls of the National Museum of Natural History for decades and inspired the artwork used in the

recent massive renovation of the fossil hall. R. Catlin, "Meet the Master Muralist Who Inspired Today's Generation of Paleoartists," *Smithsonian Magazine*, June 4, 2019, https://www.smithsonianmag.com/smithsonian-institution/Meet-master-muralist -inspired-generation-paleoartists/.

2. A. Seilacher, "Response by Adolf Seilacher," *Journal of Paleontology* 68, no. 4 (1995): 917–18.

3. A. Seilacher, *Fossil Art: An Exhibition of the Geologisches Institut Tuebingen University, Germany*, The Royal Tyrell Museum of Paleontology, 1997.

4. A detailed academic book on the long history of illustrating fossils is J. P. Davidson, *A History of Paleontology Illustration* (Bloomington: Indiana University Press, 2008). An indispensable guide for beginning and practicing paleoartists, which also reviews the history of the field, is M. P. Witton, *The Palaeoartist's Handbook* (Marlborough, UK: Crowood Press, 2018).

5. Alexander Petrunkevitch wrote monographs on fossil arachnids (spiders, scorpions, etc.). A comparison I made between his drawings and some of the original material revealed that some features that were present were not drawn and others that were missing had been added to the drawing. The prolific avocational paleontologist Erik Kjellesvig-Waering revised much of Petrunkevitch's work. He wrote me in 1977: "To say that Petrunkevitch was dated and inaccurate is probably on the kind side. Nearly 80 percent of my work was due to his careless, amateurish, and incompetent work. Yes, these are hard words. . . . But the words are true."

6. D. E. Briggs and D. Collins, "A Middle Cambrian Chelicerate from Mount Stephen, British Columbia," *Palaeontology* 31, no. 3 (1988): 779–98.

7. D. Collins, "The 'Evolution' of *Anomalocaris* and Its Classification in the Arthropod Class Dinocarida (nov.) and Order Radiodonta (nov.)," *Journal of Paleontology* 70, no. 2 (1996): 280–93.

8. M. P. Witton, D. Naish, and J. Conway, "State of the Palaeoart," *Palaeontologia Electronica* 17, no. 3 (2014), https://palaeo-electronica.org/content/2014/917-commentary -state-of-the-palaeoart.

9. *Dinosaurs Take Flight: The Art of Archaeopteryx* ran from February 2017 until May 2019.

10. The paleoartists interviewed for this chapter almost all list Gurche as an influence.

11. Apologies to all the others—I like you too!

12. L. Tapanila, J. Pruitt, A. Pradel, C. D. Wilga, J. B. Ramsay, R. Schlader, and D. A. Didier, "Jaws for a Spiral-Tooth Whorl: CT Images Reveal Novel Adaptation and Phylogeny in Fossil Helicoprion," *Biology Letters* 9, no. 2 (2013).

13. B. Page and W. Allmon, *Rock of Ages, Sands of Time: Paintings by Barbara Allmon* (Chicago: University of Chicago Press, 2001).

14. A fuller description of paleoart memes with examples, such as "The freaky giraffoid *Barosaurus* meme" was the subject of a blog by D. Naish, "Palaeoart Memes and the Unspoken Status Quo in Palaeontological Popularization," *Tetrapod Zoology* (blog), *Scientific American*, February 10, 2017, https://blogs.scientificamerican.com/tetrapod -zoology/palaeoart-memes-and-the-unspoken-status-quo-in-palaeontological -popularization/.

21. THE WORLD OF PALEONTOLOGY

1. Derek Briggs reviewed our two interpretations. See D. E. G. Briggs, "Palaeontology: How Did Eurypterids Swim?," *Nature* 320, no. 6061 (1986): 400.
2. Curt Teichert had been the editor of the *Treatise of Invertebrate Paleontology* and was an expert on fossil cephalopods. When he retired from the University of Kansas, for some inexplicable reason they kicked him out of his office. Dave Raup swooped in to give him space at the University of Rochester. He taught a memorable course on reefs that showed me how little I really knew about fossils. He was also a link to older generations of paleontologists and had a fascinating life. Curt's memoirs have recently been published, see P. M. Mikkelsen, J. T. Dutro Jr., and N. Dutro, eds., *A Paleontological Life: The Personal Memoirs of Curt Teichert* (Ithica, NY: Paleontological Research Institution, 2014).
3. This chapter draws heavily on essays written by native scientists who have firsthand knowledge of the opportunities and issues facing paleontology in their areas; I solicited them when I was editorials editor for *Palaeontologia Electronica*, under the heading "The World of Paleontology."
4. A magnificent volume on the early history of paleontology is M. J. S. Rudwick, *The Meaning of Fossils: Episodes in the History of Palaeontology* (New York: Elsevier, 1972). A second edition, which I have not read, was published in 1985. The subject was also extensively covered by Steve Gould.
5. M. Carnall, "How to Visit Lost Worlds: Our Pick of the UK's Palaeontology Museums," *The Guardian*, October 5, 2016, https://www.theguardian.com/science/2016/oct/05/palaeontology-museums-uk.
6. "postgraduate research students" in the British vernacular.
7. For a good discussion of this issue, see S. Imbler, *In London, Natural History Museums Confront Their Colonial Histories: It's Not as Easy as Returning Stolen Statues*, Atlas Obscura, October 14, 2019, https://www.atlasobscura.com/articles/decolonizing-natural-history-museum.
8. J. J. Matthews, "Perspectives: Fossils and the Law—A Summary," *Palaeontology Online* 7 (2017): 1–9, https://www.palaeontologyonline.com/articles/2017/perspectives-fossils-law-summary/.
9. R. Butler and S. Maidment, "Long-Term Career Prospects for PhD Students in Palaeontology," *Palaeontology Newsletter* 97 (2018): 42–46.
10. R. Warnock, E. Dunne, S. Giles, E. Saupe, L. Soul, and G. Lloyd, "Special Report: Are We Reaching Gender Parity Among *Palaeontology* Authors?," *Palaeontology Newsletter* 103 (2020): 40–50.
11. T. Servais, P.-O. Antoine, T. Danelian, B. Lefebvre, and B. Meyer-Berthaud, "Paleontology in France: 200 Years in the Footsteps of Cuvier and Lamarck," *Palaeontologia Electronica* 15, no. 1 (2012): 2E. This paper includes a historical survey and with a great deal of justification and claims Cuvier as the founding father of paleontology (I am sure many English will disagree).
12. W. Kiessling, A. Nützel, D. Korn, B. Kröger, and J. Müller, "German Paleontology in the Early 21st Century," *Palaeontologia Electronica* 13, no. 1 (2010): 2E.

13. Paleosynthesis project, http://www.paleosynthesis.de/.

14. The International Palaeontological Association (IPA) holds periodic International Palaeontological Congresses (https://ipc5.sciencesconf.org/) and has an official journal, *Lethaia*. This was one of the very first paleontological journals with a truly international scope.

15. A group of extinct bivalves that formed extensive reefs.

16. S. Xiao, Q. Yang, and Z.-X. Luo, "A Golden Age of Paleontology in China? A SWOT Analysis," *Palaeontologia Electronica* 13, no. 1 (2010): 3E; H. Jia, "Paleontology: Advancing China's International Leadership," *National Science Review* 6, no. 1 (2018): 171–76.

17. Q.-J. Meng, Q. Ji, Y.-G. Zhang, D. Liu, D. M. Grossnickle, and Z.-X. Luo, "An Arboreal Docodont from the Jurassic and Mammaliaform Ecological Diversification," *Science* 347, no. 6223 (2015): 764–68.

18. Geobiodiversity Database, http://www.geobiodiversity.com/home.

19. For first significant results published by GBDB, see J.-x. Fan, S.-z. Shen, D. H. Erwin, P. M. Sadler, N. MacLeod, Q.-m. Cheng, X.-d. Hou, J. Yang, X.-d. Wang, Y. Wang, H. Zhang, X. Chen, G.-x. Li, Y.-c. Zhang, Y.-k. Shi, D.-x. Yuan, Q. Chen, L.-n. Zhang, C. Li, and Y.-y. Zhao, "A High-Resolution Summary of Cambrian to Early Triassic Marine Invertebrate Biodiversity," *Science* 367, no. 6475 (2020): 272–77. For information on the Paleobiology Database (PBDB), see chapter 7.

20. D. E. G. Briggs, "Sampling the Insects of the Amber Forest," *Proceedings of the National Academy of Sciences* 115, no. 26 (2018): 6525–27.

21. H. Pohl, B. Wipfler, B. Boudinot, and R. Georg Beutel, "On the Value of Burmese Amber for Understanding Insect Evolution: Insights from †Heterobathmilla—an Exceptional Stem Group Genus of Strepsiptera (Insecta)," *Cladistics* 37, no. 2 (2020): 211–29.

22. This section is partly based on the detailed article in *Science* by Joshua Sokol. I highly recommend it. J. Sokol, "Troubled Treasure," *Science* 364, no. 6442 (2019): 722–29.

23. G. Lawton, "Blood Amber: The Exquisite Trove of Fossils Fuelling War in Myanmar," *New Scientist*, May 1, 2019, https://www.newscientist.com/article/mg24232280-600-blood-amber-the-exquisite-trove-of-fossils-fuelling-war-in-myanmar/.

24. Society for Vertebrate Paleontology, letter to editors, https://vertpaleo.org/wp-content/uploads/2021/01/SVP-Letter-to-Editors-FINAL.pdf.

25. J. D. Daza, E. L. Stanley, A. Bolet, A. M. Bauer, J. S. Arias, A. Černanský, J. J. Bevitt, P. Wagner, and S. E. Evans, "Enigmatic Amphibians in Mid-Cretaceous Amber Were Chameleon-Like Ballistic Feeders," *Science* 370, no. 6517 (2020): 687–91.

26. J. T. Haug, D. Azar, A. Ross, J. Szwedo, B. Wang, A. Arillo, V. Baranov, J. Bechteler, R. Beutel, V. Blagoderov, X. Delclòs, J. Dunlop, K. Feldberg, R. Feldmann, C. Foth, R. H. B. Fraaije, A. Gehler, D. Harms, L. Hedenäs, M. Hyžný, J. W. M. Jagt, E. A. Jagt-Yazykova, E. Jarzembowski, H. Kerp, P. K. Khine, A. G. Kirejtshuk, C. Klug, D. S. Kopylov, U. Kotthoff, J. Kriwet, R. C. McKellar, A. Nel, C. Neumann, A. Nützel, E. Peñalver, V. Perrichot, A. Pint, E. Ragazzi, L. Regalado, M. Reich, J. Rikkinen, E.-M. Sadowski, A. R. Schmidt, H. Schneider, F. R. Schram, G. Schweigert, P. Selden, L. J. Seyfullah, M. M. Solórzano-Kraemer, J. D. Stilwell, B. W. M. van Bakel, F. J. Vega, Y. Wang, L. Xing,

and C. Haug, "Comment on the Letter of the Society of Vertebrate Paleontology (SVP) Dated April 21, 2020, Regarding 'Fossils from Conflict Zones and Reproducibility of Fossil-Based Scientific Data': Myanmar Amber," *PalZ* 94, no. 3 (2020):4 31–37.

27. J. Szwedo, B. Wang, A. Soszyńska-Maj, and D. Azar, "International Palaeoentomological Society Statement," *Palaeoentomology* 3 (2020): 221–22.

28. D. E. Fernández, L. Luci, C. S. Cataldo, and D. E. Pérez, "Paleontology in Argentina: History, Heritage, Funding, and Education from a Southern Perspective," *Palaeontologia Electronica* 17, no. 3 (2014): 6E.

29. E. Gibney, "Brazil Clamps Down on Illegal Fossil Trade," *Nature* 507, no. 7490 (2014): 20.

30. R. D. O. Andrade, "Brazil Wins Fight Over Fossil Bounty," *Nature* 570 (2019): 147.

31. M. Greshko, "One-of-a-Kind Dinosaur Removed from Brazil Sparks Backlash, Investigation," *National Geographic,* December 22, 2020, https://www.nationalgeographic .com/science/2020/12/one-of-a-kind-dinosaur-removed-from-brazil-sparks-legal -investigation/.

32. J. C. Gutiérrez-Marco and D. C. García-Bellido, "The International Fossil Trade from the Paleozoic of the Anti-Atlas, Morocco," Geological Society, London, Special Publications 485. January 1, 2018, https://doi.org/10.1144/SP485.1.

33. L. Osborne, "The Fossil Frenzy," *New York Times,* October 29, 2000.

34. Department of Science and Technology, Republic of South Africa, *The South African Strategy for the Palaeosciences: Incorporating Palaeontology, Palaeo-Anthropology, and Archaeology,* https://www.gov.za/sites/default/files/gcis_document/201409/paleostrategydstfinal .pdf.

35. Centre of Excellence in Palaeosciences, University of the Witwatersrand, Johannesburg, https://www.wits.ac.za/coepalaeo/.

36. "Meet Aviwe Matiwane, a Black Female Palaeobotanist," *Women in Science,* June 23, 2020, https://www.womeninscience.africa/meet-aviwe-matiwane/.

37. J. S. Crampton and R. A. Cooper, "The State of Paleontology in New Zealand," *Palaeontologia Electronica* 13, no. 1 (2010).

38. E-mail from James Crampton, December 6, 2020.

39. C. D. Clowes, J. S. Crampton, K. J. Bland, K. S. Collins, J. G. Prebble, J. I. Raine, D. P. Strogen, M. G. Terezow, and T. Womack, "The New Zealand Fossil Record File: A Unique Database of Biological History," *New Zealand Journal of Geology and Geophysics* 64, no. 8 (2020): 1–10.

40. GNS Science (Institute of Geological and Nuclear Sciences Limited) is a descendent of the New Zealand Geological Survey.

41. Here is the full table for attendees at the 2019 NAPC, courtesy of Mary Droser: Argentina (1), Australia (12), Austria (3), Belgium (2), Brazil (6), Canada (25), Chile (1), China (31), Colombia (2), Czech Republic (2), Denmark (2), Finland (2), France (1), Germany (16), Hong Kong (6), India (6), Israel (3), Italy (1), Japan (2), Madagascar (1), Mexico (8), New Zealand (2), Norway (1), Panama (1), Poland (3), Slovakia (1), South Africa (1), South Korea (2), Spain (2), Sweden (5), Switzerland (5), Thailand (2), United Kingdom (20), United States (519).

42. For more on Pal(a)eoPERCS, visit their website at https://paleopercs.com/. The "(a)" is a nod to the British spelling.

22. LAST THOUGHTS: SWOT-ING AT PALEONTOLOGY

1. K. W. Flessa and D. M. Smith, "Paleontology in Academia: Recent Trends and Future Opportunities," Paleontology in the 21st Century (workshop), Frankfurt, Germany, 1997, http://www.nhm.ac.uk/hosted_sites/paleonet/paleo21/rr/academia.html.

INDEX

punctuated equilibria, 13, 128
pyrolysis gas chromatography mass
    spectroscopy (Py-GC-MS), 75

Rafinesquina alternata, 33–34
Raham, Gary, 258–261
range-through data, 87
Rasfeld, Rachel, 260, 263
Raup, David M., 14, 83, 96, 119, 128, 171
Raymond M. Alf Museum of Paleontology,
    62
reconstruction, producing, 256–58
reflection transformation imaging (RTI),
    69–70
Research Gate, 275
research, teaching, and service, 145–47
results, in scientific paper, 172
Rey, Luis V., 254–55, 258
Richardson, Eugene, 75, 227
RISE (Respectful Inclusive Scientific
    Events), 191
RNA, 22
rock quarries, 35
Romer, A. S., 86
Romer-Simpson Medal, 158
rotator, National Science Foundation, 199
Royal Ontario Museum (ROM), 131

Sallam, Hesham273
Santucci, Vincent, 217
scanning electron microscope (SEM), 29
scanning transmission electron
    microscope–electron energy loss
    spectroscopy (STEM-EELS), 75
Schuchert, Charles, 151–52, 230
Science Education Resource Center (SERC),
    148
Science (journal), 33, 170
ScienceBlogs, 232
science-religion divide, 247–52
Scientific American (magazine), 232
scientific paper: "A from B of C" papers,
    168; accessibility of, 168; assessing
    "quality" of, 176; broad-view type,

168–69; classes of, 167–70; disappearance
of, 175–76; example of process of,
173–75; expectation for number of, 176;
format of, 171–73; geological sciences
reflected in, 169–70; gold open access,
181; green open access, 181; H-index,
176–77; history of writing, 178–79; main
goal in, 171; new fossils, 167–68; online
publication of, 179–82; "page charges,"
179; peer review, 177–78; shifting funding
model of, 180–81
Scientists in Synagogues, 252
Scotchmoor, Judy, 59–60
Scott, Andrew, 28
Scott, Eric, 215
*Search for the Past: An Introduction to
    Paleontology* (Beerbower), 239–40
Sedimentary Geology and Paleobiology
    (SGP), National Science Foundation,
    197
Seilacher, Adolf (Dolf), 112, 172, 190, 255,
    264, 267
Selden, Paul, 264
Sepkoski, J. John, Jr. (Jack), 81, 86–87, 96,
    128, 254
Sereno, Paul, 254
serial grinding, 71–72
Servais, Thomas, 267–69
sexual harassment, 44–45, 158, 190–91
SF-METALS, 60
SF-ROCKS, 60
Sheehan, Peter, 64
Sheffield, Sarah, 241–42
Shelf Slope Experimental Taphonomy
    Initiative (SSETI), 112
shell bed, 32
shells, isotopic composition of, 105
Shelton, Sally, 63
Shimada, Kenshu, 215
Shubin, Neil, 15, 22, 31–32; fieldwork
    logistics, 37
Silurian, 6; fossils dating back to, 9–10
Simpson, George Gaylord, 40–41, 83, 272
Sinai and Synapses, 252